The Genus *Aspergillus* - Pathogenicity, Mycotoxin Production and Industrial Applications

Edited by Mehdi Razzaghi-Abyaneh and Mahendra Rai

Published in London, United Kingdom

IntechOpen

Supporting open minds since 2005

The Genus *Aspergillus* – Pathogenicity, Mycotoxin Production and Industrial Applications
http://dx.doi.org/10.5772/intechopen.94793
Edited by Mehdi Razzaghi-Abyaneh and Mahendra Rai

Contributors
Arsa Thammahong, Massimo Reverberi, Marzia Beccaccioli, Marco Zaccaria, Cameron K. K Tebbi, Ioly Kotta-Loizou, Robert H.A. Coutts, Shivaprakash M. Rudramurthy, Shreya Singh, Rimjhim Kanaujia, Mahendra Rai, Indarchand Gupta, Shital Bonde, Pramod Ingle, Sudhir Shende, Swapnil Gaikwad, Mehdi Razzaghi-Abyaneh, Aniket Gade

Notice
Statements and opinions expressed in the chapters are these of the individual contributors and not necessarily those of the editors or publisher. No responsibility is accepted for the accuracy of information contained in the published chapters. The publisher assumes no responsibility for any damage or injury to persons or property arising out of the use of any materials, instructions, methods or ideas contained in the book.

First published in London, United Kingdom, 2022 by IntechOpen
IntechOpen is the global imprint of INTECHOPEN LIMITED, registered in England and Wales, registration number: 11086078, 5 Princes Gate Court, London, SW7 2QJ, United Kingdom
Printed in Croatia

British Library Cataloguing-in-Publication Data
A catalogue record for this book is available from the British Library

Additional hard and PDF copies can be obtained from orders@intechopen.com

The Genus *Aspergillus* – Pathogenicity, Mycotoxin Production and Industrial Applications
Edited by Mehdi Razzaghi-Abyaneh and Mahendra Rai
p. cm.
Print ISBN 978-1-83968-929-1
Online ISBN 978-1-83968-930-7
eBook (PDF) ISBN 978-1-83968-931-4

Meet the editor

Professor Mehdi Razzaghi-Abyaneh obtained his Ph.D. in Mycology from Tarbiat Modares University, Tehran-Iran. In 2006/2007, he completed a training course on the identification of antifungal compounds from bioactive plants at the laboratory of the Department of Applied Biological Chemistry, Graduate School of Agriculture, Tokyo University, Japan. He is currently a full professor and head of the Mycology Department, Pasteur Institute of Iran, where he is working on mycotoxins and mycotoxigenic fungi as well as antifungal nanomaterial, and biologically active antifungals of plant, fungal, and bacterial origin. He has supervised numerous Ph.D. and MSc theses and published more than 150 papers in peer-reviewed international journals, 10 books, and several book chapters. He is studying the chemical basis of plant-fungal interactions and determining the mode of action of antifungal peptides and other small and macromolecules at cellular and molecular levels.

Professor Mahendra Rai is a visiting scientist at Nicolaus Copernicus University, Torun, Poland. He has published 380 research papers, more than 102 journal articles, and more than 60 books. He is a member of several scientific societies and has been a national scholar for five years. He received several prestigious awards, including the father T. A. Mathias award (1989) from the All India Association for Christian Higher Education, and the Medini award (1999) from the Department of Environment and Forest, Government of India. He also received a SERC visiting fellowship from the Department of Science and Technology (1996), INSA visiting fellowship from the Indian National Science Academy (1998), and TWAS-UNESCO Associateship (2002), Italy. Dr. Rai was also awarded a UGC-BSR faculty fellowship by the University Grants Commission, New Delhi. He serves as a referee for twenty international journals and is an editorial board member for ten national and international journals. He has approximately three decades of teaching and research experience. The main focus of his research is plant and nano-based bioactive against human pathogenic microbes.

Contents

Preface

The genus *Aspergillus* consists of a diverse group of airborne species with environmental and public health impacts. The members of this genus are cosmopolitan fungi frequently found in soil as the main reservoir and they are responsible for food spoilage, mycotoxin contamination, and various types of human and animal mycoses. Moreover, they are rich sources of beneficial metabolites such as antibiotics, organic acids, enzymes, and additives. At present, there are more than 350 identified species in the genus, of which around 20 species are known to be involved in the etiology of *Aspergillus*-related diseases under the common name "aspergillosis." Mode of infection is the inhalation of airborne conidia, exposure to contaminated water, and nosocomial infections. This book, which includes six chapters over two sections, discusses different aspects of the genus *Aspergillus*, including *Aspergillus*–host interactions to immunopathogenesis of aspergillosis, mycotoxin production, and industrial applications of the beneficial species.

Chapter 1 is an introductory chapter that contains useful information about all sections and chapters of the book. Chapter 2 discusses the environmental and clinical importance of *Aspergillus*–human interactions with a special focus on host immune status and previous underlying diseases as important determinants of clinical outcomes and disease spectra of aspergillosis. Chapter 3 examines the immunopathogenesis of aspergillosis with emphasis on the route of entry of etiologic *Aspergillus* species, and the function of pulmonary host defense in the clearance of infective conidia. In this context and in conditions of poor host immune response, where the neutrophils and macrophages fail to recognize the etiologic fungus, *Aspergillus* conidia attack and destroy airway epithelium. Chapter 4 reviews the role of aflatoxin in *A. flavus* resilience to stress with special attention to *Aspergillus* section *Flavi* in relation to producing the aflatoxins, secondary metabolites toxic to humans and animals. Chapter 5 examines why these fungi produce aflatoxins and the role of this mycotoxin in pathogenicity or in niche competition of producing fungus. The chapter also addresses another amazing aspect of *Aspergillus* research, the relationship between mycovirus-containing *A. flavus* and acute lymphoblastic leukemia as carcinogenesis beyond mycotoxin production. Finally, Chapter 6 discusses the industrial importance of members belonging to the genus *Aspergillus* with a focus on the ability of important species to green synthesis of functional nanomaterials with potential application in agriculture and medicine.

We would like to thank all authors for their invaluable contributions. We are also grateful to the staff at IntechOpen, especially Ms. Karmen Đaleta, who kindly assisted us in the arrangement of the book and scheduling of our activities. MR is thankful to the Polish National Agency for Academic Exchange (NAWA) for financial support (Project No. PPN/ULM/2019/1/00117/A/DRAFT/00001) to visit the Department of Microbiology, Nicolaus Copernicus University, Toruń, Poland.

Mehdi Razzaghi-Abyaneh
Department of Mycology,
Pasteur Institute of Iran,
Tehran, Iran

Mahendra Rai
Department of Microbiology,
Nicolaus Copernicus University,
Torun, Poland

Section 1

Pathogenicity

Introductory Chapter: The Genus *Aspergillus* - Pathogenicity, Mycotoxin Production and Industrial Applications

Mehdi Razzaghi-Abyaneh and Mahendra Rai

1. Introduction

Aspergillus infections in humans were firstly reported in the eighteenth century [1, 2]. *Aspergillus* was first described in 1729 by Micheli, an Italian priest and biologist, who was the first person to attempt the scientific study of fungi [3]. *A. flavus* was named and reported by Link in 1809. John Hughes Bennett (1812–1875) was the first to describe aspergillosis. In his seminal paper published in 1842, he made the very first description of *Aspergillus* growing in the lung tissue of humans [4]. Paranasal sinus mycosis in 1893 and since then numerous cases have been reported from different parts of the world. In 1926, the genus *Aspergillus* was first classified and accepted in 69 *Aspergillus* species in 11 groups. By the year 1965, the previous classification of *Aspergillus* was declared outdated, and detailed 151 species in 18 different groups were introduced. Additional research led to further refined species designations with the use of new technologies such as thin-layer chromatography of secondary metabolites and DNA hybridization.

The genus *Aspergillus* consists of numerous species gathered in a diverse group with environmental and public health importance [5, 6]. The members of this genus are cosmopolitan fungi frequently found in various natural habitats especially in soil as the main reservoir and they are responsible for food spoilage, mycotoxin contamination, and various types of human and animal mycoses [7, 8]. Moreover, they are rich sources of beneficial metabolites such as antibiotics, organic acids, enzymes, and additives. At present, there are more than 300 species which are now accepted, and new species continue to be described and added to this list. The taxonomy of species within the *Aspergillus* genus is gradually undergoing emendation with the use of molecular methods and is not yet complete. Of the known *Aspergillus* species, only 20 have been confirmed to cause human infections and three of them are consistently and regularly encountered as etiological agents of over 95% of diseases caused by members of the genus including *A. fumigatus*, *A. niger*, and *A. flavus* [9]. The other species of this genus related to human lesions are *A. terreus*, *A. glaucus*, *A. nidulans*, *A. oryzae*, and *A. clavatus*. Mode of infection is the inhalation of airborne conidia, exposure to contaminated water (contact with conidia during showering), and nosocomial infections (hospital fabrics and plastics may serve as important sources of *Aspergillus* species). The incubation period is between 2 days and 3 months.

Aspergillosis is a common term used to describe infections caused by different species of *Aspergillus* [10]. Aspergillosis was described as a clinical human disease under the name of bronchopulmonary *Aspergillus*. The species *A. fumigatus*, with *A. flavus* and *A. niger* are responsible for more than 90% of aspergillosis worldwide. A wide array of clinical forms from allergic reactions (allergic bronchopulmonary aspergillosis, rhinitis, Farmer's lung) to superficial and cutaneous infections, localized aspergilloma, and invasive infections have been reported. Invasive life-threatening aspergillosis occurs mainly in immunocompromised individuals who have undergone widespread antibiotics, cancers, or autoimmune underlying disorders. Invasive infections initiate by entering air-borne conidia to lungs with clinical entities such as invasive sinusitis, fever, facial pain, headache, cough, and dyspnea with subsequent spread to the central nervous system (CNS), leading to seizures or death.

2. Description

In the current book which comprises five distinct chapters, different aspects of the genus *Aspergillus* from *Aspergillus*-host interactions to the immunopathogenesis of aspergillosis, mycotoxin production, and industrial applications of the beneficial species have gained special attention.

It has been shown that host immune status and previous underlying diseases act as important determinants of clinical outcomes and disease spectra of aspergillosis which is life-threatening in the invasive form where the etiologic fungus affects lung tissue and disseminates to different organs with high morbidity and mortality. The role of influenza and COVID-19 infections in ICU patients has been noticed as the new risk factors of invasive aspergillosis. In relation to the immunopathogenesis of aspergillosis, documents demonstrated that following entry of causative *Aspergillus* species, fungal elements are affected by pulmonary host defense in order to clearance of infective conidia. In conditions of poor host immune response, where the neutrophils and macrophages fail to recognize the etiologic fungus, *Aspergillus* conidia attack and destroy airway epithelium and neutrophils play an important role in the clearance of fungal hyphae via oxidative and non-oxidative mechanisms. As an amazing topic in mycotoxin research, the relationship between mycovirus-containing *Aspergillus flavus* and acute lymphoblastic leukemia as carcinogenesis beyond mycotoxin production has been noticed. The role of aflatoxin in *Aspergillus flavus* resistance to stress conditions is a very interesting subject in the importance of members of *Aspergillus* section *Flavi*. In this context, it has been shown that *Aspergillus* employs a considerable amount of energy to synthesize aflatoxins which are not so obviously linked to an enhancement of population fitness. Another important aspect of the genus *Aspergillus* is the industrial application of nanomaterials produced by *Aspergillus* species. These fungi produce a large number of beneficial metabolites enabling the producing fungus to the successful synthesis of nanoparticles.

In conclusion, we would like to thank all authors for their invaluable contribution and hard work to make the successful endeavor on the goals of the present book. We are also grateful to the "In-Tech" Publisher personnel, especially Ms. Karmen Đaleta, who kindly assisted us in the arrangement of the book and scheduling our activities.

Author details

Mehdi Razzaghi-Abyaneh[1*] and Mahendra Rai[2]

1 Department of Mycology, Pasteur Institute of Iran, Tehran, Iran

2 Biotechnology Department, SGB Amravati University, Amravati, Maharashtra, India

*Address all correspondence to: mrab442@yahoo.com and mrab442@pasteur.ac.ir

IntechOpen

References

[1] Lee J. Discovery of *Aspergillus* as a Human Pathogen. http://www.antimicrobe.org/hisphoto/history/Aspergillus-Human%20Pathogens.asp

[2] Barnes EA. Short History of Invasive Aspergillosis, 1920 to 1965. The *Aspergillus* Website. (http://www.aspergillus.org.uk)

[3] Gover DW. Pier Antonio Micheli. https://www.aspergillus.org.uk/image_library/pier-antonio-micheli/?sfw=pass1651379036

[4] Bennett JH. On the parasitic vegetable structures found growing in living animals. Transact Royal Society of Edinburgh. 1842;**15**:277-294

[5] Sabino R. *Aspergillus* and health. Microorganisms. 2022;**10**(3):538

[6] Razzaghi-Abyaneh M, Chen Z-Y, Shams-Ghahfarokhi M, Rai M. Research efforts, challenges, and opportunities in mitigating aflatoxins in food and agricultural crops and its Global Health impacts. Frontiers in Microbiology. 2022;**13**:881858

[7] Razzaghi-Abyaneh M. Aflatoxins: Recent Advances and Future Prospects. Croatia: InTech Open; 2013

[8] Jamali M, Karimipour M, Shams-Ghahfarokhi M, Amani A, Razzaghi-Abyaneh M. Expression of aflatoxin genes aflO (omtB) and aflQ (ordA) differentiates levels of aflatoxin production by *aspergillus flavus* strains from soils of pistachio orchards. Research in Microbiology. 2013;**164**(4):293-299

[9] Sabino R, Veríssimo C. Novel clinical and laboratorial challenges in aspergillosis. Microorganisms. 2022;**10**(2):259

[10] Razzaghi-Abyaneh M, Shams-Ghahfarokhi M, Rai M. Medical Mycology: Current Trends and Future Prospects. Boca Raton, USA: CRC Press; 2015

Chapter 2

Aspergillus-Human Interactions: From the Environment to Clinical Significance

Arsa Thammahong

Abstract

Aspergillus species are ubiquitous fungi found in the environment worldwide. The most common *Aspergillus* species causing diseases in humans are *A. fumigatus*, *A. flavus*, *A. niger*, and *A. terreus*. However, species causing human infections are also depending on human immune status. Host immune status and previous underlying diseases are important factors leading to different clinical manifestations and different disease spectra of *Aspergillus* infections. The most severe form of *Aspergillus* infections is invasive aspergillosis in human tissue, especially invasive pulmonary aspergillosis (IPA), which has high morbidity and mortality in immunocompromised patients. ICU patients with influenza infections and COVID-19 infections are recently risk factors of invasive pulmonary aspergillosis. New diagnostic criteria include galactomannan antigen assays, nucleic acid amplification assays, and lateral flow assays for early and accurate diagnosis. Voriconazole and the newest azole, isavuconazole, are antifungals of choice in IPA. Nevertheless, azole-resistant *Aspergillus* strains are increasing throughout the world. The etiology and spreading of azole-resistant *Aspergillus* strains may originate from the widespread use of fungicides in agriculture, leading to the selective pressure of azole-resistant strains. Therefore, there is a necessity to screen *Aspergillus* antifungal susceptibility patterns for choosing an appropriate antifungal agent to treat these invasive infections. In addition, mutations in an ergosterol-producing enzyme, i.e., lanosterol 14-α demethylase, could lead to azole-resistant strains. As a result, the detection of these mutations would predict the resistance to azole agents. Although many novel azole agents have been developed for invasive *Aspergillus* infections, the rate of novel antifungal discovery is still limited. Therefore, better diagnostic criteria and extensive antifungal resistant *Aspergillus* screening would guide us to better manage invasive *Aspergillus* infections with our existing limited resources.

Keywords: *Aspergillus*, *Aspergillus*-human interactions, invasive aspergillosis, antifungal susceptibility test, azole, voriconazole, amphotericin B, influenza-associated pulmonary aspergillosis, COVID-19-associated pulmonary aspergillosis

1. Introduction

Aspergillus species are saprophytic ubiquitous filamentous fungi [1]. They are in Phylum Ascomycota with both sexual and asexual forms [1]. In their sexual form, they produce asci and ascospores within the appropriate environment, while they produce conidia, or asexual spores, on phialides surrounding their vesicles at the tip of conidiophores in their asexual form [1]. *Aspergillus* conidia are different in size and shape depending on *Aspergillus* species, which affects the dispersion and infectivity properties of *Aspergillus* [1]. Their conidia can be found in the soil, decomposed piles, air, animals, and humans. They cause diseases in immunocompromised hosts, e.g., patients with acquired immunodeficiency syndrome (AIDS), allogenic hematopoietic stem cell transplant or solid organ transplant candidates, patients with immunosuppressive drugs, patients with prolonged neutropenia, and patients with other underlying diseases [2]. The common pathogenic *Aspergillus* species are *A. fumigatus, A. flavus, A. niger,* and *A. terreus* [3]. There are a wide variety of disease spectra of *Aspergillus* infections, i.e., invasive aspergillosis, chronic aspergillosis, and allergic forms of aspergillosis [1, 2]. The most severe form causing high morbidity and mortality rate, especially in immunocompromised hosts, is invasive aspergillosis (IA) [2, 4]. An increase of immunocompromised hosts would also increase patients with IA with a high mortality rate [4–14].

Invasive aspergillosis (IA) is recently increasing in patients with allogenic hematopoietic stem cell transplantation (HSCT) and solid organ transplantation [5, 8, 13, 15–22]. Underlying conditions of patients with IA are hematological malignancies, e.g., leukemia or lymphoma, bone marrow transplant, and solid-organ transplant patients [5, 8, 13, 15–22]. Recently, not only neutropenic patients are at risk for IA, but non-neutropenic patients with immunosuppressive agents, e.g., biologics, small-molecule kinase inhibitors (SMKIs), Chimeric Antigen Receptor (CAR) T cells, are also at risk [23–28]. In developing countries, poor-controlled diabetes mellitus is one of the critical risk factors of IA [10, 12]. Therefore, risk factors of IA are now patients with malignancy, autoimmune, inflammatory diseases, complex immune-metabolic diseases from aging, immunosuppressive treatment, previous septic conditions, novel biologic treatment, including patients with hematological malignancies receiving SMKIs, patients in ICU, patients with a cytokine storm syndrome from CAR-T cells treated with high-dose corticosteroids, patients in ICU with severe influenza or other viral infections [23–36]. In an era of Coronavirus Disease 2019 (COVID-19) infections, IA was recognized as a severe complication of patients with COVID-19 infections in ICU [37–46].

2. Pathogenesis of *Aspergillus* and its virulence factors

Among thousands of *Aspergillus* species, only less than twenty species could cause diseases in humans [47]. The pathogenic species usually possess virulence factors that help them survive and cause infections inside hosts. *Aspergillus fumigatus* was utilized as a model to study virulence factors in many studies (**Table 1**) [1].

To survive inside the host environment, *Aspergillus* species need to adapt to heat and hypoxic conditions inside hosts. For the heat stress, the trehalose pathway was shown to have a role in heat tolerance and virulence of *A. fumigatus* [47]. Heat shock proteins (HSPs), especially Hsp90, are chaperone proteins associated with stress tolerance, not only for heat [48–50]. In mammalians, HIF1α, as a common transcription factor, controls cellular homeostasis in hypoxic conditions [51]. In fungi, a homolog of HIF1α, called the sterol regulatory element-binding protein (SREBP) or SrbA in *A. fumigatus,* is induced by hypoxia and iron starvation conditions [52–56]. SrbA protein is also associated with the virulence of *A. fumigatus in vivo* [52–54].

A. fumigatus possesses enzymes to protect itself against host reactive oxygen species (ROS), e.g., catalase, superoxide dismutases, thioredoxin, glutathione, including mitochondrial electron transport chain [57–62]. In some animal

Virulence factors	Characteristics
Stress tolerance	• Thermotolerance
	• Hypoxic adaptation
	• pH/Reactive oxygen species (ROS) resistance
	• Secondary metabolites
	• Light response
Metabolism and nutrient uptake	• siderophores, Zinc Magnesium Copper transporter, calmodulin, calcineurin, phosphate permeases
Cell components	• Cell wall: β-glucan, chitin, rodlet
	• Galactosaminogalactan (GAG)
	• Melanin
Others	• Biofilm
	• Cellular heterogeneity

Table 1.
Essential virulence factors in Aspergillus fumigatus requiring for causing infections inside humans [1].

models, e.g., an eye infection model, demonstrated that these fungal enzymes were essential for fungal virulence [63]. Secondary metabolites are also playing a role in fungal virulence [64–66]. *A. fumigatus* secondary metabolites are gliotoxin, fumigaclavine, trypacidin, helvolic acid, fumitremorgin, fumagillin, and pseurotin, associated with host cellular toxicity [67–71]. However, the mechanisms behind this toxicity is still unclear and need to be further investigated *in vivo* [71]. *A. flavus* produces aflatoxins, which are important carcinogenic secondary metabolites, and other secondary metabolites, called Velvet complex, as environmental response mechanisms [72, 73]. Circadian rhythms or light response, which were studied thoroughly in the *Neurospora* model system, are essential to react with the environment [74]. Light-induced mycelial pigmentation and germination acted as a stress signaling pathway in *A. fumigatus* via transcription factor LreA and FphA, respectively [75–77].

For nutrient acquisition, exoenzymes or proteases are major enzymes produced by *A. fumigatus,* especially the alkaline protease Alp1 and the metalloprotease Mep1 [1, 78]. In *A. fumigatus,* a transcriptional repressor called CreA has a vital role in carbon catabolite repression. *Af*CreA regulates growth on different nitrogen, carbon, and lipid sources and has a role in amino acid transportation, nitrogen, and carbon assimilation, including glycogen and trehalose metabolism [79, 80]. Although CreA is not required for virulence, it is required for disease progression in invasive pulmonary aspergillosis (IPA) mouse models [79–81]. For nitrogen utilization, *Af*RhbA, a Ras-related protein in a nitrogen-regulated signaling pathway, and *Af*AreA, a GATA transcription factor requiring the expression of genes involving nitrogen utilization, are related to virulence in *A. fumigatus* [82–84]. *A. fumigatus* still needs divalent cations, i.e., iron, copper, magnesium, zinc, calcium, for its growth and virulence inside hosts via siderophores, calmodulin, calcineurin, specific importers, and exporters [85, 86].

Additionally, cell wall components of *Aspergillus fumigatus* are also essential virulence factors for fungal survival inside hosts and are important for host immune response [87–92]. Cell wall components consist of β-1,3-glucan, chitin, galactomannan, α-1,3-glucan, and melanin depending on different stages of *A. fumigatus,* i.e., conidial, or hyphal stage [91–95]. β-1,3-glucan, a central component of *Aspergillus* cell wall polysaccharide, is a pathogen-associated molecular pattern (PAMP) recognized by host pattern recognition receptors (PRR), e.g., dectin-1 [88]. During its conidial stage, rodlet, or hydrophobins, and dihydroxynapthalene (DHN) melanin are present to protect fungal conidia against host immune response by evading host pathogen-associated molecular patterns (PAMPs) recognition, including protecting fungi from unfavorable stress conditions [93–97]. Furthermore, in its hyphal stage, galactosaminogalactan (GAG), which is a water-insoluble polymer consisting of a pyranose-form galactose, galactosamine, and N-acetylgalactosamine (GalNAc), is present as an extracellular matrix on an outer layer of the cell wall [98]. GAG is associated with biofilm formation and immunosuppression properties by masking PAMP exposure and resisting neutrophil killing via neutrophil extracellular traps (NETs) [99–102]. The linkage between cell wall components and metabolic pathways is still unclear. Nevertheless, these components share the same building blocks, e.g., UDP-glucose, glucose 6-phosphate, with specific metabolic pathways, e.g., glycolysis, trehalose biosynthesis pathway [81, 103–105]. It is possible that the homeostasis of cell wall biosynthesis is involved with some metabolic pathways, e.g., the trehalose biosynthesis pathway. Disruption of one of these trehalose enzymes or building blocks would result in decreased virulence due to changes in cell wall compositions [81, 103–105]. Understanding this homeostasis would lead to the discovery of novel antifungal targets in the future.

3. Diagnosis of invasive *Aspergillus* infections: challenge in the field

Aspergillus infections are associated closely with host immune status [106, 107]. Severe asthma with fungal sensitization and allergic bronchial pulmonary aspergillosis (ABPA) are found in immunocompetent hosts with hypersensitivity, while aspergilloma and chronic pulmonary aspergillosis are found in immunocompetent hosts with previous structural diseases, such as lung cavity from previous tuberculosis infections [108]. In immunocompromised hosts, invasive aspergillosis is common and severe, causing high morbidity and mortality in patients [108, 109].

For invasive pulmonary aspergillosis, early diagnosis and prompt treatment are the keys to decrease the disease burden. Differentiation between *Aspergillus* colonization and invasive infections is still challenging [25, 92, 93]. Recently, the revised EORTC guideline for diagnosis of invasive fungal infections, including *Aspergillus* infections, recommended the diagnostic criteria including host factors, clinical, radiological, and microbiological criteria with new diagnostic methods (**Table 2**) [109]. Proven invasive aspergillosis is confirmed with histopathologic, cytopathologic, microscopic analysis, or nucleic acid analysis of sterile specimens or tissue or formalin-fixed paraffin-embedded tissue (FFPE), including culture recovered from sterile sites [109]. Species of common *Aspergillus* recovered from cultures are differentiated using macroscopic and microscopic morphology, but the nucleic acid analysis is necessary for the species complex (**Table 3**) [110]. For probable and possible invasive aspergillosis, host factors, clinical features, and mycological evidence are including for the diagnosis of invasive aspergillosis. Host factors include the history of neutropenia, which is less than 500 neutrophils/mm^3, for more than ten days, hematological malignancy, allogenic stem cell transplantation, solid organ transplantation, therapeutic-dose corticosteroids at not less than 0.3 mg/kg for not less than three weeks during the previous 60 days, treatment with T-cell or B-cell immunosuppressants, inherited immunodeficiency, or acute graft-versus-host disease grade III or IV [109]. For clinical evidence of pulmonary aspergillosis, a chest high-resolution CT scan is recommended to observe any halo

Diagnosis of invasive aspergillosis.	Criteria
Proven	• Microscopic analysis: from needle aspiration or biopsy OR
	• Culture: from sterile sites except for BAL fluid, paranasal sinuses, and urine OR
	• Tissue nucleic acid analysis from formalin-fixed paraffin-embedded tissue
Probable: 1 host factor + 1 clinical feature+1 mycological evidence	Host factors
	• Recent neutropenia
Possible: 1 host factor + 1 clinical feature	• Hematological malignancy
	• Receipt of an allogenic stem cell transplant
	• Receipt of a solid organ transplant
	• Prolonged use of corticosteroids
	• Use of T-cell immunosuppressants
	• Use of B-cell immunosuppressants
	• Inherited severe immunodeficiency
	• Acute GVHD grade III or IV
	Clinical features: pulmonary aspergillosis
	• One of the following CT Chest patterns:
	○ Dense well-circumscribed lesion with or without a halo sign
	○ Air crescent sign
	○ Cavity
	○ Wedge-shaped and segmental or lobar consolidation
	Mycological evidence
	• Culture positive from sputum, BAL, bronchial brush, or aspirate
	• Direct examination positive from sputum, BAL, bronchial brush, or aspirate
	• Galactomannan antigen: plasma serum BAL CSF: any of:
	○ Single serum or plasma >/= 1
	○ BAL fluid >/= 1
	○ Single serum or plasma >/= 0.7 and BAL fluid >/= 0.8
	○ CSF >/= 1
	• *Aspergillus* PCR: any of:
	○ Plasma, serum, or whole blood 2 or more consecutive PCR
	○ BAL fluid 2 or more duplicate PCR
	○ At least 1 PCR from plasma serum or whole blood & 1 PCR from BAL fluid

Table 2.
Diagnosis of invasive aspergillus infections from revised EORTC/MSG criteria 2020 (BAL: bronchoalveolar lavage; CT: computed tomography; CSF: cerebrospinal fluid; GVHD: graft versus host disease; PCR: polymerase chain reaction) [109].

sign, air-crescent sign, cavity, or wedge-shaped and segmental or lobar consolidation [109, 111]. Probable invasive aspergillosis still needs at least one mycological evidence to support the diagnosis. Mycological evidence is including cultures recovered from sputum, bronchoalveolar lavage (BAL), bronchial brush, or

Aspergillus species	Macroscopic features	Microscopic features
Aspergillus fumigatus	Typical blue-green colony with suede-like surface	Columnar uniseriate conidial heads with phialides limited to upper two-thirds of its vesicles; short and smooth conidiophores; basipetal green, rough-walled globose to subglobose conidia
Aspergillus flavus	Bright to dark yellow-green colony with a granular, flat surface	Radiate biseriate conidial heads with phialides over the surface of mature vesicles; coarsely rough conidiophores; pale green, globose to subglobose conidia
Aspergillus niger	Dark brown to the black colony with white to yellow color at the reverse side of the colony	Globose, large, dark brown, biseriate, radiate conidial head with long metulae; smooth, hyaline conidiophores; dark brown, rough conidia
Aspergillus terreus	Cinnamon-brown colony with suede-like surface and yellow to deep brown color at the reverse side of the colony	Compact, columnar, biseriate conidial heads; hyaline, smooth conidiophores; hyaline to yellow, smooth-walled conidia

Table 3.
Macroscopic and microscopic features of clinical-relevant Aspergillus species (colony on Czapek Dox agar at 30°C) [110].

aspirate [109]. *Aspergillus* galactomannan antigen assays with different thresholds depending on specimens, including serum, BAL fluid, plasma, and cerebrospinal fluid (CSF), support the diagnosis of invasive aspergillosis [112–115]. However, decreased sensitivity of galactomannan antigen assay is observed in patients with anti-mold therapy [115]. In addition, *Aspergillus* PCR from blood and BAL fluid is introduced to confirm the diagnosis and identify specific *Aspergillus* species with certain mutations related to triazole resistance [109, 116–124].

Nonetheless, revised EORTC/MSG criteria for diagnosing invasive fungal infections may be applied mainly for neutropenic patients or immunocompromised patients. Therefore, specific guidelines for the diagnosis of invasive aspergillosis in non-neutropenic patients in ICU (Invasive pulmonary aspergillosis in ICU, AspICU) or patients with influenza (Influenza-associated pulmonary aspergillosis, IAPA) or Severe Acute Respiratory Syndrome Coronavirus-2 (SARS-CoV-2) (COVID-19 associated pulmonary aspergillosis, CAPA) co-infections were developed and published for early and accurate diagnosis (**Table 4**) [31, 125–127].

4. Treatment of *Aspergillus* infections

IA also includes the infections of the lower respiratory system, sinuses, and skin as entry routes. In addition, the cardiovascular system, central nervous system, and other tissues could be infected from hematogenous dissemination or direct extension from adjacent infected tissues [2]. Infectious Diseases Society of America (IDSA, 2016) and ESCMID-ECMM-ERS (2017) recommended voriconazole (6 mg/ kg, intravenous route every 12 hours for one day, and then 4 mg/kg every 12 hours; 200–300 mg every 12-hour, oral route) as a first-line treatment for invasive pulmonary aspergillosis (IPA) [2, 128]. For alternative treatment, liposomal amphotericin B (3–5 mg/kg/day, intravenous route) and isavuconazole (200 mg every 8 hours for three days and then 200 mg daily) [2]. For other invasive aspergillosis syndromes, i.e., invasive sinus aspergillosis, tracheobronchial aspergillosis, invasive aspergillosis of the central nervous system or cardiovascular system, *Aspergillus* osteomyelitis,

Diagnostic criteria of IPA	AspICU [125]	IPA with influenza (IAPA) [126]	IPA with SARS-CoV-2 (CAPA) [127]
Host factors	One of the following: • Neutropenia (<500/mm^3) before or at ICU admission • Hematological or oncological malignancy with cytotoxic therapy • Glucocorticoid treatment with prednisolone equivalent >20 mg/day • Immunodeficiency	Entry criteria: influenza-like illness + positive influenza PCR or antigen + timing (7 days before and 96 hours after ICU admission)	Entry criteria: patients with COVID-19 infections (RT-PCR) in ICU with a temporal relationship to suspected IPA
Clinical features	One of the following: • Fever with appropriate antibiotic treatment for at least three days • Recurrent fever after a fever-free period for at least 48 hours with antibiotics and without other apparent cause • Dyspnea • Hemoptysis • Pleuritic chest pain or pleural friction rub • Worsening respiratory failure with appropriate antibiotics and ventilator support	None	None
Radiological evidence	• Any medical imaging by conventional chest X-ray or CT scan of lungs	• Pulmonary infiltrate OR • Cavitating infiltrate (not from other causes)	• Pulmonary infiltrate OR • Cavitating infiltrate (not from other causes)

Diagnostic criteria of IPA	AspICU [125]	IPA with influenza (IAPA) [126]	IPA with SARS-CoV-2 (CAPA) [127]

Microbiological evidence	• *Aspergillus* recovered from the lower respiratory tract (LRT) (entry criterion)
	• *Aspergillus*-positive culture of BAL fluid without bacterial growth together with a positive microscopic analysis showing branching hyphae (if no host factor)
		• If pulmonary infiltrate presents, at least one of the following:
		  ○ Galactomannan (GM) antigen assay: serum >0.5 or BAL ≥ 1.0 or
		  ○ positive culture from BAL
		• If lung cavity presents, at least one of the following: positive sputum culture or tracheal aspirate culture
			Probable CAPA: at least one of the following:
			• Microscopic detection of septate hyphae in BAL
			• Positive BAL culture
			• Serum GM >0.5 or serum LFA index >0.5
			• BAL GM ≥1.0 or BAL LFA index ≥1.0
			• Two or more positive *Aspergillus* PCR in plasma, serum, or whole blood or a single positive *Aspergillus* PCR in BAL (<36 cycles); or a single positive *Aspergillus* PCR in plasma, serum, or whole blood with a single positive in BAL fluid (any threshold cycle)
			Possible CAPA: at least one of the following:
			• Microscopic detection of septate hyphae in non-BAL
			• Positive non-BAL culture
			• Single non-BAL GM >4·5
			• Non-BAL GM >1·2 twice or more
			• Non-BAL GM >1·2 plus another non-BAL PCR or LFA positive

Diagnostic criteria of IPA	AspICU [125]	IPA with influenza (IAPA) [126]	IPA with SARS-CoV-2 (CAPA) [127]
Categories	• Proven IPA: similar to EORTC/MSG 2020 criteria • Putative IPA: *Aspergillus*-positive from LRT + Clinical evidence + Radiological evidence + (Host factors or *Aspergillus* culture from BAL with positive microscopic analysis) • Colonization: ≥ 1 criterion for a diagnosis of putative IPA is not fulfilled	• Proven IAPA: entry criteria with tissue diagnosis similar to EORTC/MSG 2020 criteria • Putative IAPA: entry criteria + Radiological evidence + Microbiological evidence • Colonization: ≥ 1 criterion for a diagnosis of putative IPA is not fulfilled	• Proven CAPA: entry criteria with tissue diagnosis similar to EORTC/MSG 2020 criteria • Probable CAPA: entry criteria + radiological evidence + probable criteria of microbiological evidence • Possible CAPA: entry criteria + radiological evidence + possible criteria of microbiological evidence

Table 4.
Diagnostic criteria for invasive pulmonary aspergillosis (IPA) of patients in ICU (AspICU) or with influenza (IAPA) or SARS-CoV-2 (CAPA) coinfections (PCR: polymerase chain reaction; ICU: intensive care unit; RT-PCR: Real-time polymerase chain reaction; BAL: bronchoalveolar lavage; GM: galactomannan; LFA: lateral flow assay) [31, 125–127].

Aspergillus endophthalmitis and keratitis, cutaneous aspergillosis, and *Aspergillus* peritonitis, intravenous voriconazole is still the first-line therapy [2]. For IPA in ICU patients, patients with hematological malignancies, or solid organ transplants, IAPA, and CAPA, voriconazole and isavuconazole are still recommended as the first-line treatment (**Table 5**).

Voriconazole is metabolized at the liver via CYP2C19 and CYP3A4 [135]. Medications with CYP2C19 and CYP3A4-dependent metabolism, antacids, proton pump inhibitors may affect serum voriconazole levels [136]. Adverse reactions and toxicity of voriconazole are associated with higher serum voriconazole levels [137]. Common adverse reactions include reversible visual disturbances, hepatotoxicity, photosensitivity, reversible visual or auditory hallucinations, tachyarrhythmias, and QT interval prolongations [137, 138]. Isavuconazole is a second-generation broad-spectrum triazole requiring a loading dose with a five-day half-life [139]. Isavuconazole has fewer adverse reactions in photosensitivity, hepatotoxicity, visual abnormality, and less drug–drug interaction [140–142]. Isavuconazole is a CYP3A4 inhibitor and can decrease the metabolism of sirolimus, tacrolimus, cyclosporine, and digoxin, leading to increased levels of these agents [142]. Furthermore, isavuconazole can induce dose-dependent QTc shortening [143]. Isavuconazole was shown to be non-inferior to voriconazole to treat invasive mold disease from the SECURE trial [144]. Posaconazole is also a broad-spectrum triazole used mainly for prophylaxis and salvage treatment of invasive fungal infections [145]. A suspension form of posaconazole has unpredictable bioavailability and needs a high-fat meal for better absorption [146]. However, tablet and IV formulations overcome this limitation. Posaconazole strongly inhibits CYP3A4 and is metabolized through UGT1A4 [145]. Using CYP3A4 substrates with posaconazole should be cautious [145]. The common adverse effects of posaconazole are gastrointestinal disturbances, hepatotoxicity, rashes, fever, hypokalemia, hypomagnesemia, and QTc prolongation [145].

Amphotericin B, a polyene antifungal agent binding to ergosterol in the fungal cell membrane, has many forms, i.e., conventional with deoxycholate and lipid-based form [2, 147]. Conventional amphotericin B has common adverse effects, including acute reactions after infusion (fever, chills, nausea), phlebitis, hypokalemia, hypomagnesemia, and nephrotoxicity (usually from renal tubular acidosis). The lipid-based form has less nephrotoxicity than the conventional form [2]. Nevertheless, acute infusion reactions may still present in liposomal amphotericin B [148]. In addition, hypokalemia, hypomagnesemia, mild bilirubin, alkaline phosphatase elevations are also present occasionally in lipid-based amphotericin B [2]. Lipid-based amphotericin B is recommended for alternative treatment of invasive aspergillosis in case that azoles cannot be used [2].

Echinocandins, e.g., caspofungin, micafungin, is a non-competitive β-1,3 D-glucan synthase inhibitor leading to loss of fungal cell wall's strength and integrity [149, 150]. Echinocandins have fewer adverse reactions and fewer drug–drug interactions [149, 150]. Echinocandins are recommended for salvage therapy or in azole-resistant *Aspergillus* infections combined with azoles for invasive aspergillosis treatment (**Table 5**) [2, 151–153].

Therapeutic drug monitoring (TDM) of azoles, e.g., voriconazole, posaconazole, isavuconazole, is necessary, especially in elderly patients, obese patients, critically ill patients, and patients with potential azole drug–drug interactions [2]. For treatment of IA, IDSA recommended TDM of voriconazole at a trough level of >1–1.5 µg/mL but less than 5–6 µg/mL to prevent neurotoxicity [2]. American Society of Transplantation Infection Diseases Community of Practice (AST) recommended TDM of posaconazole (suspension and tablet form) and isavuconazole at a trough level of >1–1.25 µg/mL and 2–3 µg/mL, respectively [154]. Timing

Condition	First-line treatment	Prophylaxis
IPA in ICU patients [129, 130]	Voriconazole (6 mg/kg, intravenous route every 12 hours for one day, and then 4 mg/kg every 12 hours; 200–300 mg every 12 hours oral route) or Isavuconazole (200 mg every 8 hours for 3 days and then 200 mg daily) (Liposomal amphotericin B, 3–5 mg/kg/day, intravenous route, in ICU patients with severe liver insufficiency, cirrhosis Child-Pugh scores B, C)	In immunocompetent patients in ICU, prophylaxis is not recommended for IPA
IPA in patients with hematological malignancies [131–133]	Voriconazole or Isavuconazole (Liposomal amphotericin B as alternative treatment)	Posaconazole (oral solution 200 mg every eight hours or tablet/intravenous route 300 mg every 12 h on day one then 300 mg daily) (in AML and MDS undergoing intensive chemotherapy with the incidence of invasive mold diseases >8% or in graft-versus-host disease) Voriconazole (200 mg orally every 12 h) (in HSCT)
IPA in patients with solid organ transplantation [134]	Voriconazole or Isavuconazole (Liposomal amphotericin B in hepatotoxicity, drug–drug interaction, ≥10% environment azole-resistant isolates found)	Kidney and heart transplantation are not recommended Lung transplantation: voriconazole, nebulized liposomal amphotericin B
IAPA [31, 126]	Voriconazole or Isavuconazole	No current recommendation Need further studies
CAPA [127]	For azole sensitive: Voriconazole or Isavuconazole (for 6–12 weeks) For azole-resistant: - Suspected: voriconazole + echinocandin (Caspofungin 70 mg first day followed by 50 mg daily) or isavuconazole + echinocandin - Proven: Liposomal amphotericin B	No current recommendation Need further studies

Table 5.
Treatment of invasive pulmonary aspergillosis (IPA) in ICU patients, patients with hematological malignancies, or solid organ transplants, influenza-associated pulmonary aspergillosis (IAPA), and COVID-19 associated pulmonary aspergillosis (CAPA) (AML: acute myeloid leukemia; MDS: myelodysplastic syndrome; HSCT: hematological stem cell transplantation).

for measuring serum trough concentration of voriconazole, posaconazole, and isavuconazole is at 5–7 days, after 5 days, and within 7 days, respectively [154]. For prophylaxis, International Society for Heart and Lung Transplantation (ISHLT) recommended TDM of voriconazole and posaconazole at a trough level of ≥1 μg/mL and > 0.7 μg/mL, respectively [155]. Additionally, in CAPA, ECMM/ISHAM recommended weekly TDM of voriconazole and posaconazole at a trough level of 2–6 µg/mL and 1–3.75 µg/mL, respectively [127].

5. Azole-resistant *Aspergillus*

5.1 Etiology and clinical significance

Voriconazole and isavuconazole are the first-line therapy of invasive aspergillosis [2, 129, 130]. Furthermore, azoles, i.e., posaconazole and voriconazole, are also used as prophylaxis of invasive aspergillosis in patients with hematological malignancies and solid organ transplantation [131–134]. Therefore, azoles are important antifungal agents to combat invasive aspergillosis. Unfortunately, azole-resistant *Aspergillus fumigatus* strains are emerging and increasing, leading to increased treatment failure and mortality [156, 157]. The etiology of these emerging azole-resistant *A. fumigatus* (ARAF) may be from the environmental selective pressure associated with azole fungicides in the agricultural area, including Europe, Asia, Latin America, the Midwest, and Southeast states of the USA [158–161]. The supporting evidence of environment-derived ARAF is that ARAF strains were recovered from azole-naive patients [158, 162–165]. In addition, the most common mutations at the *cyp51A* gene (encoding lanosterol 14-α demethylase) causing azole resistance in ARAF strains, which are TR$_{34}$/L98H and TR$_{46}$/Y121F/T289A mutations, were also recovered from patients' homes and surroundings [166–171].

Azole fungicides, i.e., bromuconazole, difenoconazole, epoxiconazole, enilconazole, metconazole, prochloraz, propiconazole, prothioconazole-desthio, and tebuconazole, play an important role in the development of environment-derived azole-resistant *Aspergillus* isolates leading to cross resistance to medical azoles [169, 172, 173].

Antifungal susceptibility tests (AST) of *Aspergillus* species are essential for screening azole-resistant *Aspergillus* isolates. The indications to perform *Aspergillus* AST are that the fungus is recovered from sterile sites in regions with high azole-resistant rates, including long-term azole treatment in chronic bronchopulmonary aspergillosis and breakthrough *Aspergillus* infections or recurrent or persistent infections [2, 128, 174].

The standard antifungal susceptibility testing of filamentous fungi to observe the minimum inhibitory concentration (MIC) using broth microdilution assays was described by the Clinical and Laboratory Standards Institute (CLSI) and the European Union Committee on Antimicrobial Susceptibility Testing (EUCAST) [175, 176]. To determine antifungal resistance of *Aspergillus* species, e.g., *A. flavus*, *A. fumigatus*, *A. niger*, *A. terreus*, CLSI and EUCAST utilized two values, which are epidemiological cutoff values (ECVs or ECOFFs) and clinical breakpoints (BP) (**Table 6**). ECVs for CLSI and ECOFFs for EUCAST of each antifungal agent against each *Aspergillus* originate from MIC distribution of the wild-type *Aspergillus* population [175–178]. These values can divide *Aspergillus* strains into two groups, which are wild-type and non-wild-type strains. Non-wild-type strains may resist those antifungal agents [175, 176, 178]. Clinical breakpoints are based on antifungal pharmacodynamics, pharmacokinetics, data from clinical trials, and patient outcomes [175, 176, 178]. Resistance is determined by the MICs over R (resistant) (**Table 6**). For EUCAST, another value is the area of technical uncertainty (ATU), which is the value that needs to be addressed before reporting these results, i.e., repeating the test, using a genotypic test, changing the susceptibility category, or including ATU as a part of the report [176].

Aspergillus species	Antifungal agents	CLSI M59 & M61, 2020 (µg/mL)				EUCAST BP_ECOFF v2.0, 2020 (µg/mL)			
		ECV	S	I	R	ECV	S≤	R>	ATU
A. flavus	Amphotericin B	4	—	—	—	4	—	—	—
	Caspofungin	0.5	—	—	—	—	—	—	—
	Isavuconazole	1	—	—	—	2	1	2	2
	Itraconazole	1	—	—	—	1	1	1	2
	Posaconazole	0.5	—	—	—	0.5	—	—	—
	Voriconazole	2	—	—	—	2	—	—	—
A. fumigatus	Amphotericin B	2	—	—	—	1	1	1	—
	Caspofungin	0.5	—	—	—	—	—	—	—
	Isavuconazole	1	—	—	—	2	1	2	2
	Itraconazole	1	—	—	—	1	1	1	2
	Posaconazole	—	—	—	—	0.25	0.125	0.25	0.25
	Voriconazole	1	≤0.5	1	≥2	1	1	1	2
A. niger	Amphotericin B	2	—	—	—	0.5	1	1	—
	Caspofungin	0.25	—	—	—	—	—	—	—
	Isavuconazole	4	—	—	—	4	—	—	—
	Itraconazole	4	—	—	—	4	—	—	—
	Posaconazole	2	—	—	—	0.5	—	—	—
	Voriconazole	2	—	—	—	2	—	—	—
A. terreus	Amphotericin B	4	—	—	—	8	—	—	—
	Caspofungin	0.12	—	—	—	—	—	—	—
	Isavuconazole	1	—	—	—	1	1	1	—
	Itraconazole	2	—	—	—	0.5	1	1	2
	Posaconazole	1	—	—	—	0.25	0.125	0.25	0.25
	Voriconazole	2	—	—	—	2	—	—	—

Table 6.
Interpretation of antifungal susceptibility tests and epidemiological cutoff values (ECVs) of Aspergillus species according to CLSI M59 and M61, 2020 and EUCAST BP ECOFF version 2, 2020 (S: susceptible, I: intermediate, R: resistant, ATU: Area of Technical Uncertainty) [175–177].

Molecular methods to detect *CYP51A* mutations, e.g., TR$_{34}$/L98H, TR$_{46}$/Y121F, are established by using classic PCRs with sequencing, real-time PCRs, loop-mediated isothermal amplification (LAMP), or whole-genome sequencing (WGS) [179]. These molecular methods have a high negative predictive value to rule out these resistant strains' infections [179]. However, they had narrow coverage and mutations at this point depending on association data between mutations and antifungal resistance property. Furthermore, commercial tools are still not approved by the US FDA [179].

5.2 Management of azole-resistant *Aspergillus* and novel antifungal candidates

Overexpression with a tandem repeat in the promoter area (TR$_{34}$ or TR$_{46}$) and point mutations (L98H or Y121F/T289A) in the *cyp51A* gene, encoding azole's target called lanosterol 14-α demethylase, would lead to azole resistance in *Aspergillus fumigatus* including voriconazole and isavuconazole [156, 178]. To treat these azole-resistant *Aspergillus* infections, monotherapy of each azole should be avoided, especially in areas with more than 10% of azole resistance prevalence [180]. In areas with high rates of azole resistance, liposomal amphotericin B and a combination of voriconazole and echinocandin should be considered [2, 127, 128, 156, 180]. Therefore, the prevalence of azole-resistant *Aspergillus* strains using conventional culturing methods together with broth microdilution assays or using molecular biology (RT-PCR) is essential to decide the optimal treatment and to choose suitable antifungal agents to get rid of these infections [156, 179].

From the increased speed of azole-resistant *Aspergillus* strains, novel antifungal agents with high efficacy and fewer side effects are crucial to combat these infections with very high mortality [156]. However, discovering these novel antifungal agents has many steps and methods to evaluate both *in vitro* and *in vivo* analyses for both antifungal activity and toxicity [181, 182]. The first step for screening antifungal activity has many methods depending on the screening purpose [181]. To observe the antifungal activity of novel antifungal candidates, the broth microdilution method is the standard method to provide the MICs [183]. This method is perfect for various compounds requiring high throughput assays [181]. Furthermore, this method requires a small number of compounds and can apply to different *Aspergillus* species simultaneously [181]. To observe combinatorial effects between novel antifungal candidates and current antifungal agents, checkerboard assays are used to determine the fractional inhibitory concentration index (FICI) [184, 185]. The FICI is calculated using the sum of the fractional inhibitory concentration (FIC_1) of the first compound, which is MIC_{1+2} of the combination of the first and the second compounds divided by MIC_1 of the first compound alone, and the FIC_2 of the second compound [184, 185]. Synergistic, additive, indifferent, and antagonistic effects are defined by FICI ≤ 0.5; >0.5–1; >1–4; and >4, respectively [184–186]. For the cytotoxicity effects on human epithelial cells, many *in vitro* colorimetric assays, including mammalian tissue culture systems and vital dyes, are used, such as Alamar blue, MTT, XTT (tetrazolium) assays [181]. Next steps after *in vitro* studies to prove the antifungal activity and toxicity, *in vivo* animal models are used to study pharmacodynamics and pharmacokinetics, including *in vivo* antifungal activity and *in vivo* toxicity [181]. Then, these antifungal candidates would follow through the clinical trial phase I (safety), phase II (checking effectiveness), phase III (confirming effectiveness, side effects), and get approved [181, 182, 187].

Many novel antifungal compounds against both classical targets and novel targets are in clinical trials (**Table** 7) [262]. Novel targets against *Aspergillus* species include glycosylphosphatidylinositol (GPI) anchor protein, dihydroorotate dehydrogenase in pyrimidine synthesis, fungal mitochondrial respiration chain, siderophore iron transporter, Heat shock protein 90 (Hsp90), calcineurin, histone deacetylase (HDAC), inositol phosphorylceramide (IPC) synthase, chitin synthase, and sphingolipid pathway (**Table** 7). Nevertheless, more clinical trials are on the way for these agents before using them in the clinical practice against antifungal-resistant *Aspergillus*/fungal strains.

In addition, enzymes in the *Aspergillus* trehalose biosynthesis pathway, i.e., trehalose-6-phosphate synthase, trehalose-6-phosphate phosphatase, trehalase enzymes, were identified as important virulence factors, including proteins related to the trehalose pathway, i.e., *Af*SsdA, *Af*TslA [103, 105, 263, 264]. The trehalose pathway in *A. fumigatus* is associated with cell wall integrity and fungal virulence *in vivo* [103, 264, 265]. However, inhibitors of this pathway are still lacking and under-investigated. Validamycin A is one of the inhibitors of trehalase enzymes and was first demonstrated its strong antifungal activity against a plant fungal

Name.	Target	Mechanism	Advantage	Administration	Clinical trial
Classical targets					
Encochleated amphotericin B (CAmB) [188–192]	Ergosterol	Renovated structure of amphotericin B with cochleated lipid-crystal nanoparticles	Oral administration, broad-spectrum, less toxicity	Oral	Phase I
Rezafungin (CD101) [192–203]	1,3-β-D-glucan synthase (FKS)	1,3-β-D-glucan synthase inhibitor	Improved stability, long half-life (once a week), activity against A. fumigatus, A. terreus, A. flavus, and A. niger	Intravenous	Phase III
Ibrexafungerp (SCY-078) [204–210]	1,3-β-D-glucan synthase (FKS)	1,3-β-D-glucan synthase inhibitor	activity against A. fumigatus, A. terreus, A. flavus, and A. niger; including itraconazole-resistant Aspergillus	Oral and intravenous	Phase III
VT-1598 [211, 212]	Lanosterol demethylase (CYP51)	Tetrazole, inhibiting lanosterol demethylase	Less drug–drug interactions, long half-life, broad-spectrum: Candida, Aspergillus	Oral	Phase I
VT-1161 (oteseconazole) [213, 214]	Lanosterol demethylase (CYP51)	Tetrazole, inhibiting lanosterol demethylase	Less drug–drug interactions, long half-life: activity against azole-resistant Candida, onychomycosis	Oral	Phase III
VT-1129 (quilseconazole) [215–220]	Lanosterol demethylase (CYP51)	Tetrazole, inhibiting lanosterol demethylase	Less drug–drug interactions, long half-life, brain penetration, activity against Cryptococcus, Candida	Oral	Phase I
PC945 [221–227]	Lanosterol demethylase (CYP51)	Triazole, inhibiting lanosterol demethylase	Fungicidal, high lung exposure, activity against A. fumigatus	Inhalation	Phase II
Novel targets					
Fosmanogepix (APX001) [228–236]	Glycosylphosphatidylinositol (GPI) anchor protein synthesis (GWT1)	Inhibiting GPI	Fungal-specific target, broad-spectrum, activity against A. fumigatus, A. terreus, A. flavus, and A. niger	Intravenous and oral	Phase II
APX2096 [236]	Glycosylphosphatidylinositol (GPI) anchor protein synthesis (GWT1)	Inhibiting GPI	Strong activity against Cryptococcus, effective blood–brain barrier penetration	Intraperitoneal and oral	—
Olorofim (F901318) [237–239]	Dihydroorotate dehydrogenase in pyrimidine synthesis	Inhibiting pyrimidine synthesis	Activity against A. fumigatus, A. terreus, A. flavus, and A. nidulans, including azole-resistant A. fumigatus	Intravenous and oral	Phase III

Name.	Target	Mechanism	Advantage	Administration	Clinical trial
T-2307 [240–242]	Intracellular mitochondrial membrane respiration potential	Inhibiting mitochondrial respiration chain (arylamidine)	Uptaking more by fungal cells, fungicidal activity against *A. fumigatus, A. terreus, A. flavus, A. nidulans,* and *A. niger*	Subcutaneous	Phase I
VL–2397 (ASP2397) [243–245]	Unknown	Uptaking by siderophore iron transporter (Sit1)	Fungicidal, activity against *A. fumigatus, A. terreus, A. flavus,* and *A. niger*	Intravenous	Phase II
Geldanamycin [246–248]	Heat shock protein 90 (Hsp90)	Inhibiting Hsp90	Synergy to caspofungin	Intravenous	—
Tacrolimus (FK506) [249–251]	Calcineurin	Inhibiting calcineurin	Synergy to caspofungin, activity against *A. fumigatus*	Intravenous and oral	—
Cyclosporin A [249, 252]	Calcineurin	Inhibiting calcineurin	Activity against *A. fumigatus*	Intravenous, oral, and topical	—
FK506 analogs (9D31OD-FK506) [251]	Calcineurin	Inhibiting calcineurin	Synergy to azoles, decrease T-cell toxicity and host immunosuppression	Intravenous	—
Trichostatin A [253]	Histone deacetylase (HDAC)	Inhibiting HDAC	Synergy to caspofungin, activity against *A. fumigatus*	Intravenous	—
MGCD290 [254]	Histone deacetylase (HDAC)	Inhibiting HDAC	Synergy to caspofungin, azole, broad spectrum	Oral	Phase II
Aureobasidin A [255–258]	Inositol phosphorylceramide (IPC) synthase	Inhibiting IPC synthase	Synergy to caspofungin	Intravenous and oral	—
Nikkomycin [259]	Chitin synthase	Inhibiting chitin synthase	Broad-spectrum	Intravenous	—
BHBM D13 [260, 261]	Sphingolipid pathway	Acylhydrazone, inhibiting fungal sphingolipid glucosylceramide (GlcCer) synthesis	Broad-spectrum, specific to fungi, fungicidal, blood–brain barrier penetration, less toxicity	Intraperitoneal and oral	—

Table 7.
Summary of novel antifungal agents against classical targets and novel targets for Aspergillus infections.

pathogen, *Rhizoctonia solani* [266–269]. Furthermore, validamycin A has antifungal activity against *Candida albicans* and *Aspergillus flavus* [186, 270]. Validamycin A also possesses combinatorial effects with conventional amphotericin B against *A. flavus* [186]. Nevertheless, *in vivo* experiments are still necessary to verify an antifungal activity of validamycin A. Additionally, the high-osmolarity glycerol (HOG)-mitogen-activated protein kinase (MAPK) signaling pathway is associated with trehalose production and stress response in *A. fumigatus* [271–274]. This signaling pathway may be another good antifungal target to be developed in the future. Therefore, there are many more pathways involved with *Aspergillus* virulence, and there are so many unexplored areas in *Aspergillus* pathogenesis to develop novel antifungal candidates. With this knowledge, we could overcome the shortage of antifungal agents against many more antifungal-resistant *Aspergillus* strains to emerge very soon.

6. Conclusion

Aspergillus species are common fungi found everywhere around humans. They adapt and express many virulence factors to survive inside hosts and cause infections in immunocompromised hosts. Recently, new risk factors that cause severe invasive pulmonary aspergillosis are ICU patients with influenza infections or COVID-19 infections. The diagnosis of invasive aspergillosis, especially without proven tissue or culture evidence, is still challenging. New molecular methods, i.e., nucleic acid assays, lateral flow assays, are introduced for supporting the diagnosis of probable and possible invasive aspergillosis. Nevertheless, voriconazole and isavuconazole are the first-line therapy in IPA in ICU patients, patients with hematological malignancies, patients with IAPA, and CAPA. Furthermore, posaconazole is the principal antifungal agent for the prophylactic treatment of IPA in patients with hematological malignancies. Additionally, emerging azole-resistant *Aspergillus* strains are increasing, and the management against these azole-resistant *Aspergillus* strains is the combination therapy between azoles and echinocandins, including liposomal amphotericin B. Although novel antifungal agents against *Aspergillus* species are on their way, antimicrobial stewardship of existing antifungal agents is also crucial to prevent further breakthrough antifungal-resistant strains in the future. With our better understanding of *Aspergillus* pathogenesis, the shortage of antifungal agents against *Aspergillus* and its resistant strains would no longer be for the better lives of patients suffering from *Aspergillus* infections.

Acknowledgements

The author would like to thank the Department of Microbiology, Faculty of Medicine, Chulalongkorn University, Bangkok and Bamrasnaradura Infectious Diseases Institute (BIDI), Department of Disease Control, Ministry of Public Health, Nonthaburi, Thailand for all their supports.

Conflict of interest

The author declares no conflict of interest.

Author details

Arsa Thammahong
Department of Microbiology, Faculty of Medicine, Chulalongkorn University, Bangkok, Thailand

*Address all correspondence to: arsa.t@chula.ac.th

IntechOpen

References

[1] Latge JP, Chamilos G. *Aspergillus fumigatus* and Aspergillosis in 2019. Clin Microbiol Rev. 2019;33(1).

[2] Patterson TF, Thompson GR, 3rd, Denning DW, Fishman JA, Hadley S, Herbrecht R, et al. Practice Guidelines for the Diagnosis and Management of Aspergillosis: 2016 Update by the Infectious Diseases Society of America. Clin Infect Dis. 2016;63(4):e1-e60.

[3] Sugui JA, Kwon-Chung KJ, Juvvadi PR, Latge JP, Steinbach WJ. *Aspergillus fumigatus* and related species. Cold Spring Harb Perspect Med. 2014;5(2):a019786.

[4] Brown GD, Denning DW, Gow NA, Levitz SM, Netea MG, White TC. Hidden killers: human fungal infections. Sci Transl Med. 2012;4(165):165rv13.

[5] Fracchiolla NS, Sciume M, Orofino N, Guidotti F, Grancini A, Cavalca F, et al. Epidemiology and treatment approaches in management of invasive fungal infections in hematological malignancies: Results from a single-centre study. PLoS One. 2019;14(5):e0216715.

[6] Slavin MA, Chakrabarti A. Opportunistic fungal infections in the Asia-Pacific region. Med Mycol. 2012;50(1):18-25.

[7] Kriengkauykiat J, Ito JI, Dadwal SS. Epidemiology and treatment approaches in management of invasive fungal infections. Clin Epidemiol. 2011;3: 175-91.

[8] Pfaller MA, Diekema DJ. Epidemiology of invasive mycoses in North America. Crit Rev Microbiol. 2010;36(1):1-53.

[9] Lehrnbecher T, Frank C, Engels K, Kriener S, Groll AH, Schwabe D. Trends in the postmortem epidemiology of invasive fungal infections at a university hospital. J Infect. 2010;61(3):259-65.

[10] Chakrabarti A, Chatterjee SS, Das A, Shivaprakash MR. Invasive aspergillosis in developing countries. Med Mycol. 2011;49 Suppl 1:S35-47.

[11] Chakrabarti A, Chatterjee SS, Shivaprakash MR. Overview of opportunistic fungal infections in India. Nihon Ishinkin Gakkai Zasshi. 2008;49(3):165-72.

[12] Thammahong A, Thayidathara P, Suksawat K, Chindamporn A. Epidemiology of invasive Aspergillosis in a tertiary-care hospital of Thailand, 2006-2011. Mycoses. 2012;55:230-.

[13] Graf K, Khani SM, Ott E, Mattner F, Gastmeier P, Sohr D, et al. Five-years surveillance of invasive aspergillosis in a university hospital. BMC Infect Dis. 2011;11:163.

[14] Gangneux JP, Camus C, Philippe B. Epidemiology of invasive aspergillosis and risk factors in non neutropaenic patients. Rev Mal Respir. 2010;27(8): e34-46.

[15] Nucci M, Queiroz-Telles F, Tobon AM, Restrepo A, Colombo AL. Epidemiology of opportunistic fungal infections in Latin America. Clin Infect Dis. 2010;51(5):561-70.

[16] Neofytos D, Fishman JA, Horn D, Anaissie E, Chang CH, Olyaei A, et al. Epidemiology and outcome of invasive fungal infections in solid organ transplant recipients. Transpl Infect Dis. 2010;12(3):220-9.

[17] Kontoyiannis DP, Marr KA, Park BJ, Alexander BD, Anaissie EJ, Walsh TJ, et al. Prospective surveillance for invasive fungal infections in hematopoietic stem cell transplant recipients, 2001-2006: overview of the Transplant-Associated

Infection Surveillance Network
(TRANSNET) Database. Clin Infect Dis.
2010;50(8):1091-100.

[18] Pappas PG, Alexander BD,
Andes DR, Hadley S, Kauffman CA,
Freifeld A, et al. Invasive fungal
infections among organ transplant
recipients: results of the Transplant-
Associated Infection Surveillance
Network (TRANSNET). Clin Infect Dis.
2010;50(8):1101-11.

[19] Neofytos D, Horn D, Anaissie E,
Steinbach W, Olyaei A, Fishman J, et al.
Epidemiology and outcome of invasive
fungal infection in adult hematopoietic
stem cell transplant recipients: analysis
of Multicenter Prospective Antifungal
Therapy (PATH) Alliance registry. Clin
Infect Dis. 2009;48(3):265-73.

[20] Azie N, Neofytos D, Pfaller M,
Meier-Kriesche HU, Quan SP, Horn D.
The PATH (Prospective Antifungal
Therapy) Alliance(R) registry and
invasive fungal infections: update 2012.
Diagnostic microbiology and infectious
disease. 2012;73(4):293-300.

[21] Robin C, Cordonnier C, Sitbon K,
Raus N, Lortholary O, Maury S, et al.
Mainly Post-Transplant Factors Are
Associated with Invasive Aspergillosis
after Allogeneic Stem Cell
Transplantation: A Study from the
Surveillance des Aspergilloses Invasives
en France and Societe Francophone de
Greffe de Moelle et de Therapie
Cellulaire. Biol Blood Marrow
Transplant. 2019;25(2):354-61.

[22] Siopi M, Karakatsanis S,
Roumpakis C, Korantanis K,
Sambatakou H, Sipsas NV, et al. A
Prospective Multicenter Cohort
Surveillance Study of Invasive
Aspergillosis in Patients with
Hematologic Malignancies in Greece:
Impact of the Revised EORTC/MSGERC
2020 Criteria. J Fungi (Basel). 2021;7(1).

[23] Herbrecht R, Bories P, Moulin JC,
Ledoux MP, Letscher-Bru V. Risk

stratification for invasive aspergillosis in
immunocompromised patients. Ann N
Y Acad Sci. 2012;1272:23-30.

[24] Ghez D, Calleja A, Protin C,
Baron M, Ledoux MP, Damaj G, et al.
Early-onset invasive aspergillosis and
other fungal infections in patients
treated with ibrutinib. Blood.
2018;131(17):1955-9.

[25] Chamilos G, Lionakis MS,
Kontoyiannis DP. Call for Action:
Invasive Fungal Infections Associated
With Ibrutinib and Other Small
Molecule Kinase Inhibitors Targeting
Immune Signaling Pathways. Clin Infect
Dis. 2018;66(1):140-8.

[26] Bazaz R, Denning DW. Subacute
Invasive Aspergillosis Associated With
Sorafenib Therapy for Hepatocellular
Carcinoma. Clin Infect Dis. 2018;67(1):
156-7.

[27] Hill JA, Li D, Hay KA, Green ML,
Cherian S, Chen X, et al. Infectious
complications of CD19-targeted
chimeric antigen receptor-modified
T-cell immunotherapy. Blood.
2018;131(1):121-30.

[28] Park JH, Romero FA, Taur Y,
Sadelain M, Brentjens RJ, Hohl TM,
et al. Cytokine Release Syndrome Grade
as a Predictive Marker for Infections in
Patients With Relapsed or Refractory
B-Cell Acute Lymphoblastic Leukemia
Treated With Chimeric Antigen
Receptor T Cells. Clin Infect Dis.
2018;67(4):533-40.

[29] Benjamim CF, Lundy SK,
Lukacs NW, Hogaboam CM, Kunkel SL.
Reversal of long-term sepsis-induced
immunosuppression by dendritic cells.
Blood. 2005;105(9):3588-95.

[30] Taccone FS, Van den Abeele AM,
Bulpa P, Misset B, Meersseman W,
Cardoso T, et al. Epidemiology of
invasive aspergillosis in critically ill
patients: clinical presentation,

underlying conditions, and outcomes. Crit Care. 2015;19:7.

[31] Schauwvlieghe A, Rijnders BJA, Philips N, Verwijs R, Vanderbeke L, Van Tienen C, et al. Invasive aspergillosis in patients admitted to the intensive care unit with severe influenza: a retrospective cohort study. Lancet Respir Med. 2018;6(10):782-92.

[32] Huang L, Zhang Y, Hua L, Zhan Q. Diagnostic value of galactomannan test in non-immunocompromised critically ill patients with influenza-associated aspergillosis: data from three consecutive influenza seasons. Eur J Clin Microbiol Infect Dis. 2021.

[33] Waldeck F, Boroli F, Suh N, Wendel Garcia PD, Flury D, Notter J, et al. Influenza-associated aspergillosis in critically-ill patients-a retrospective bicentric cohort study. Eur J Clin Microbiol Infect Dis. 2020;39(10): 1915-23.

[34] van de Veerdonk FL, Kolwijck E, Lestrade PP, Hodiamont CJ, Rijnders BJ, van Paassen J, et al. Influenza-Associated Aspergillosis in Critically Ill Patients. Am J Respir Crit Care Med. 2017;196(4):524-7.

[35] Cornillet A, Camus C, Nimubona S, Gandemer V, Tattevin P, Belleguic C, et al. Comparison of epidemiological, clinical, and biological features of invasive aspergillosis in neutropenic and nonneutropenic patients: a 6-year survey. Clin Infect Dis. 2006;43(5): 577-84.

[36] Jenks JD, Nam HH, Hoenigl M. Invasive aspergillosis in critically ill patients: Review of definitions and diagnostic approaches. Mycoses. 2021.

[37] Arastehfar A, Carvalho A, van de Veerdonk FL, Jenks JD, Koehler P, Krause R, et al. COVID-19 Associated Pulmonary Aspergillosis (CAPA)-From Immunology to Treatment. J Fungi (Basel). 2020;6(2).

[38] Mohamed A, Rogers TR, Talento AF. COVID-19 Associated Invasive Pulmonary Aspergillosis: Diagnostic and Therapeutic Challenges. J Fungi (Basel). 2020;6(3).

[39] Lai CC, Yu WL. COVID-19 associated with pulmonary aspergillosis: A literature review. J Microbiol Immunol Infect. 2021;54(1):46-53.

[40] Apostolopoulou A, Esquer Garrigos Z, Vijayvargiya P, Lerner AH, Farmakiotis D. Invasive Pulmonary Aspergillosis in Patients with SARS-CoV-2 Infection: A Systematic Review of the Literature. Diagnostics (Basel). 2020;10(10).

[41] Marr KA, Platt A, Tornheim JA, Zhang SX, Datta K, Cardozo C, et al. Aspergillosis Complicating Severe Coronavirus Disease. Emerg Infect Dis. 2021;27(1).

[42] Machado M, Valerio M, Alvarez-Uria A, Olmedo M, Veintimilla C, Padilla B, et al. Invasive pulmonary aspergillosis in the COVID-19 era: An expected new entity. Mycoses. 2021;64(2):132-43.

[43] Costantini C, van de Veerdonk FL, Romani L. Covid-19-Associated Pulmonary Aspergillosis: The Other Side of the Coin. Vaccines (Basel). 2020;8(4).

[44] Koehler P, Bassetti M, Chakrabarti A, Chen SCA, Colombo AL, Hoenigl M, et al. Defining and managing COVID-19-associated pulmonary aspergillosis: the 2020 ECMM/ISHAM consensus criteria for research and clinical guidance. Lancet Infect Dis. 2020.

[45] Chong WH, Neu KP. The Incidence, Diagnosis, and Outcomes of COVID-19-associated Pulmonary Aspergillosis (CAPA): A Systematic Review. J Hosp Infect. 2021.

[46] Mitaka H, Kuno T, Takagi H, Patrawalla P. Incidence and Mortality of COVID-19-associated Pulmonary Aspergillosis: A Systematic Review and Meta-analysis. Mycoses. 2021.

[47] Hohl TM, Feldmesser M. *Aspergillus fumigatus*: principles of pathogenesis and host defense. Eukaryot Cell. 2007;6(11):1953-63.

[48] O'Meara TR, Cowen LE. Hsp90-dependent regulatory circuitry controlling temperature-dependent fungal development and virulence. Cell Microbiol. 2014;16(4):473-81.

[49] Robbins N, Uppuluri P, Nett J, Rajendran R, Ramage G, Lopez-Ribot JL, et al. Hsp90 governs dispersion and drug resistance of fungal biofilms. PLoS Pathog. 2011;7(9): e1002257.

[50] Schneider A, Blatzer M, Posch W, Schubert R, Lass-Florl C, Schmidt S, et al. *Aspergillus fumigatus* responds to natural killer (NK) cells with upregulation of stress related genes and inhibits the immunoregulatory function of NK cells. Oncotarget. 2016;7(44): 71062-71.

[51] Friedrich D, Fecher RA, Rupp J, Deepe GS, Jr. Impact of HIF-1alpha and hypoxia on fungal growth characteristics and fungal immunity. Microbes Infect. 2017;19(3):204-9.

[52] Barker BM, Kroll K, Vodisch M, Mazurie A, Kniemeyer O, Cramer RA. Transcriptomic and proteomic analyses of the *Aspergillus fumigatus* hypoxia response using an oxygen-controlled fermenter. BMC genomics. 2012;13:62.

[53] Blatzer M, Barker BM, Willger SD, Beckmann N, Blosser SJ, Cornish EJ, et al. SREBP coordinates iron and ergosterol homeostasis to mediate triazole drug and hypoxia responses in the human fungal pathogen *Aspergillus fumigatus*. PLoS Genet. 2011;7(12): e1002374.

[54] Chung D, Barker BM, Carey CC, Merriman B, Werner ER, Lechner BE, et al. ChIP-seq and in vivo transcriptome analyses of the *Aspergillus fumigatus* SREBP SrbA reveals a new regulator of the fungal hypoxia response and virulence. PLoS Pathog. 2014;10(11): e1004487.

[55] Losada L, Barker BM, Pakala S, Pakala S, Joardar V, Zafar N, et al. Large-scale transcriptional response to hypoxia in *Aspergillus fumigatus* observed using RNAseq identifies a novel hypoxia regulated ncRNA. Mycopathologia. 2014;178(5-6):331-9.

[56] Vodisch M, Scherlach K, Winkler R, Hertweck C, Braun HP, Roth M, et al. Analysis of the *Aspergillus fumigatus* proteome reveals metabolic changes and the activation of the pseurotin A biosynthesis gene cluster in response to hypoxia. J Proteome Res. 2011;10(5): 2508-24.

[57] Shibuya K, Paris S, Ando T, Nakayama H, Hatori T, Latge JP. Catalases of *Aspergillus fumigatus* and inflammation in aspergillosis. Nihon Ishinkin Gakkai Zasshi. 2006;47(4): 249-55.

[58] Paris S, Wysong D, Debeaupuis JP, Shibuya K, Philippe B, Diamond RD, et al. Catalases of *Aspergillus fumigatus*. Infect Immun. 2003;71(6):3551-62.

[59] Lambou K, Lamarre C, Beau R, Dufour N, Latge JP. Functional analysis of the superoxide dismutase family in *Aspergillus fumigatus*. Mol Microbiol. 2010;75(4):910-23.

[60] Kurucz V, Kruger T, Antal K, Dietl AM, Haas H, Pocsi I, et al. Additional oxidative stress reroutes the global response of *Aspergillus fumigatus* to iron depletion. BMC genomics. 2018;19(1):357.

[61] Burns C, Geraghty R, Neville C, Murphy A, Kavanagh K, Doyle S.

Identification, cloning, and functional expression of three glutathione transferase genes from *Aspergillus fumigatus*. Fungal Genet Biol. 2005;42(4):319-27.

[62] Grahl N, Dinamarco TM, Willger SD, Goldman GH, Cramer RA. *Aspergillus fumigatus* mitochondrial electron transport chain mediates oxidative stress homeostasis, hypoxia responses and fungal pathogenesis. Mol Microbiol. 2012;84(2):383-99.

[63] Leal SM, Jr., Vareechon C, Cowden S, Cobb BA, Latge JP, Momany M, et al. Fungal antioxidant pathways promote survival against neutrophils during infection. J Clin Invest. 2012;122(7):2482-98.

[64] Macheleidt J, Mattern DJ, Fischer J, Netzker T, Weber J, Schroeckh V, et al. Regulation and Role of Fungal Secondary Metabolites. Annu Rev Genet. 2016;50:371-92.

[65] Valiante V. The Cell Wall Integrity Signaling Pathway and Its Involvement in Secondary Metabolite Production. J Fungi (Basel). 2017;3(4).

[66] Raffa N, Keller NP. A call to arms: Mustering secondary metabolites for success and survival of an opportunistic pathogen. PLoS Pathog. 2019;15(4): e1007606.

[67] Amitani R, Taylor G, Elezis EN, Llewellyn-Jones C, Mitchell J, Kuze F, et al. Purification and characterization of factors produced by *Aspergillus fumigatus* which affect human ciliated respiratory epithelium. Infect Immun. 1995;63(9):3266-71.

[68] Sugui JA, Pardo J, Chang YC, Zarember KA, Nardone G, Galvez EM, et al. Gliotoxin is a virulence factor of *Aspergillus fumigatus*: gliP deletion attenuates virulence in mice immunosuppressed with hydrocortisone. Eukaryot Cell. 2007;6(9):1562-9.

[69] Spikes S, Xu R, Nguyen CK, Chamilos G, Kontoyiannis DP, Jacobson RH, et al. Gliotoxin production in *Aspergillus fumigatus* contributes to host-specific differences in virulence. J Infect Dis. 2008;197(3):479-86.

[70] Scharf DH, Heinekamp T, Remme N, Hortschansky P, Brakhage AA, Hertweck C. Biosynthesis and function of gliotoxin in *Aspergillus fumigatus*. Appl Microbiol Biotechnol. 2012;93(2):467-72.

[71] Brown R, Priest E, Naglik JR, Richardson JP. Fungal Toxins and Host Immune Responses. Frontiers in microbiology. 2021;12:643639.

[72] Amare MG, Keller NP. Molecular mechanisms of *Aspergillus flavus* secondary metabolism and development. Fungal Genet Biol. 2014;66:11-8.

[73] Amaike S, Keller NP. *Aspergillus flavus*. Annu Rev Phytopathol. 2011;49:107-33.

[74] Fuller KK, Dunlap JC, Loros JJ. Light-regulated promoters for tunable, temporal, and affordable control of fungal gene expression. Appl Microbiol Biotechnol. 2018;102(9):3849-63.

[75] Fuller KK, Ringelberg CS, Loros JJ, Dunlap JC. The fungal pathogen *Aspergillus fumigatus* regulates growth, metabolism, and stress resistance in response to light. mBio. 2013;4(2).

[76] Fuller KK, Cramer RA, Zegans ME, Dunlap JC, Loros JJ. *Aspergillus fumigatus* Photobiology Illuminates the Marked Heterogeneity between Isolates. mBio. 2016;7(5).

[77] Chen S, Fuller KK, Dunlap JC, Loros JJ. Circadian Clearance of a Fungal Pathogen from the Lung Is Not Based on Cell-intrinsic Macrophage Rhythms. J Biol Rhythms. 2018;33(1): 99-105.

[78] Sriranganadane D, Waridel P, Salamin K, Reichard U, Grouzmann E, Neuhaus JM, et al. *Aspergillus* protein degradation pathways with different secreted protease sets at neutral and acidic pH. J Proteome Res. 2010;9(7):3511-9.

[79] Ries LN, Beattie SR, Espeso EA, Cramer RA, Goldman GH. Diverse Regulation of the CreA Carbon Catabolite Repressor in *Aspergillus nidulans*. Genetics. 2016;203(1):335-52.

[80] Beattie SR, Mark KMK, Thammahong A, Ries LNA, Dhingra S, Caffrey-Carr AK, et al. Filamentous fungal carbon catabolite repression supports metabolic plasticity and stress responses essential for disease progression. PLoS Pathog. 2017;13(4): e1006340.

[81] de Assis LJ, Manfiolli A, Mattos E, Fabri J, Malavazi I, Jacobsen ID, et al. Protein Kinase A and High-Osmolarity Glycerol Response Pathways Cooperatively Control Cell Wall Carbohydrate Mobilization in *Aspergillus fumigatus*. mBio. 2018;9(6).

[82] Panepinto JC, Oliver BG, Fortwendel JR, Smith DL, Askew DS, Rhodes JC. Deletion of the *Aspergillus fumigatus* gene encoding the Ras-related protein RhbA reduces virulence in a model of Invasive pulmonary aspergillosis. Infect Immun. 2003;71(5):2819-26.

[83] Dietl AM, Amich J, Leal S, Beckmann N, Binder U, Beilhack A, et al. Histidine biosynthesis plays a crucial role in metal homeostasis and virulence of *Aspergillus fumigatus*. Virulence. 2016;7(4):465-76.

[84] Hensel M, Arst HN, Jr., Aufauvre-Brown A, Holden DW. The role of the *Aspergillus fumigatus* areA gene in invasive pulmonary aspergillosis. Molecular & general genetics: MGG. 1998;258(5):553-7.

[85] Blatzer M, Latge JP. Metal-homeostasis in the pathobiology of the opportunistic human fungal pathogen *Aspergillus fumigatus*. Curr Opin Microbiol. 2017;40:152-9.

[86] Fleck CB, Schobel F, Brock M. Nutrient acquisition by pathogenic fungi: nutrient availability, pathway regulation, and differences in substrate utilization. Int J Med Microbiol. 2011;301(5):400-7.

[87] Zacharias CA, Sheppard DC. The role of *Aspergillus fumigatus* polysaccharides in host-pathogen interactions. Curr Opin Microbiol. 2019;52:20-6.

[88] Hatinguais R, Willment JA, Brown GD. PAMPs of the Fungal Cell Wall and Mammalian PRRs. Curr Top Microbiol Immunol. 2020;425:187-223.

[89] Ahamefula Osibe D, Lei S, Wang B, Jin C, Fang W. Cell wall polysaccharides from pathogenic fungi for diagnosis of fungal infectious disease. Mycoses. 2020;63(7):644-52.

[90] Fontaine T, Latge JP. Galactomannan Produced by *Aspergillus fumigatus*: An Update on the Structure, Biosynthesis and Biological Functions of an Emblematic Fungal Biomarker. J Fungi (Basel). 2020;6(4).

[91] Beauvais A, Latge JP. Special Issue: Fungal Cell Wall. J Fungi (Basel). 2018;4(3).

[92] Gow NAR, Latge JP, Munro CA. The Fungal Cell Wall: Structure, Biosynthesis, and Function. Microbiol Spectr. 2017;5(3).

[93] Valsecchi I, Lai JI, Stephen-Victor E, Pille A, Beaussart A, Lo V, et al. Assembly and disassembly of *Aspergillus fumigatus* conidial rodlets. Cell Surf. 2019;5:100023.

[94] Valsecchi I, Dupres V, Stephen-Victor E, Guijarro JI, Gibbons J, Beau R, et al. Role of Hydrophobins in *Aspergillus fumigatus*. J Fungi (Basel). 2017;4(1).

[95] Valsecchi I, Dupres V, Michel JP, Duchateau M, Matondo M, Chamilos G, et al. The puzzling construction of the conidial outer layer of *Aspergillus fumigatus*. Cell Microbiol. 2019;21(5):e12994.

[96] Tsai HF, Wheeler MH, Chang YC, Kwon-Chung KJ. A developmentally regulated gene cluster involved in conidial pigment biosynthesis in *Aspergillus fumigatus*. J Bacteriol. 1999;181(20):6469-77.

[97] Bayry J, Beaussart A, Dufrene YF, Sharma M, Bansal K, Kniemeyer O, et al. Surface structure characterization of *Aspergillus fumigatus* conidia mutated in the melanin synthesis pathway and their human cellular immune response. Infect Immun. 2014;82(8):3141-53.

[98] Fontaine T, Delangle A, Simenel C, Coddeville B, van Vliet SJ, van Kooyk Y, et al. Galactosaminogalactan, a new immunosuppressive polysaccharide of *Aspergillus fumigatus*. PLoS Pathog. 2011;7(11):e1002372.

[99] Briard B, Muszkieta L, Latge JP, Fontaine T. Galactosaminogalactan of *Aspergillus fumigatus*, a bioactive fungal polymer. Mycologia. 2016;108(3):572-80.

[100] Gravelat FN, Beauvais A, Liu H, Lee MJ, Snarr BD, Chen D, et al. *Aspergillus* galactosaminogalactan mediates adherence to host constituents and conceals hyphal beta-glucan from the immune system. PLoS Pathog. 2013;9(8):e1003575.

[101] Lee MJ, Geller AM, Bamford NC, Liu H, Gravelat FN, Snarr BD, et al. Deacetylation of Fungal Exopolysaccharide Mediates Adhesion and Biofilm Formation. mBio. 2016;7(2):e00252-16.

[102] Lee MJ, Liu H, Barker BM, Snarr BD, Gravelat FN, Al Abdallah Q, et al. The Fungal Exopolysaccharide Galactosaminogalactan Mediates Virulence by Enhancing Resistance to Neutrophil Extracellular Traps. PLoS Pathog. 2015;11(10):e1005187.

[103] Thammahong A, Caffrey-Card AK, Dhingra S, Obar JJ, Cramer RA. *Aspergillus fumigatus* Trehalose-Regulatory Subunit Homolog Moonlights To Mediate Cell Wall Homeostasis through Modulation of Chitin Synthase Activity. mBio. 2017;8(2).

[104] Thammahong A, Puttikamonkul S, Perfect JR, Brennan RG, Cramer RA. Central Role of the Trehalose Biosynthesis Pathway in the Pathogenesis of Human Fungal Infections: Opportunities and Challenges for Therapeutic Development. Microbiol Mol Biol Rev. 2017;81(2).

[105] Thammahong A, Dhingra S, Bultman KM, Kerkaert JD, Cramer RA. An Ssd1 Homolog Impacts Trehalose and Chitin Biosynthesis and Contributes to Virulence in *Aspergillus fumigatus*. mSphere. 2019;4(3).

[106] Pirofski LA, Casadevall A. The damage-response framework of microbial pathogenesis and infectious diseases. Adv Exp Med Biol. 2008;635:135-46.

[107] Park SJ, Mehrad B. Innate immunity to *Aspergillus* species. Clin Microbiol Rev. 2009;22(4):535-51.

[108] Moldoveanu B, Gearhart AM, Jalil BA, Saad M, Guardiola JJ. Pulmonary Aspergillosis: Spectrum of Disease. Am J Med Sci. 2021;361(4): 411-9.

[109] Donnelly JP, Chen SC, Kauffman CA, Steinbach WJ, Baddley JW, Verweij PE, et al. Revision and Update of the Consensus Definitions of Invasive Fungal Disease From the European Organization for Research and Treatment of Cancer and the Mycoses Study Group Education and Research Consortium. Clin Infect Dis. 2020;71(6):1367-76.

[110] Walsh TJ, Hayden RT, Larone DH. Larone's medically important fungi: a guide to identification2018.

[111] Park SY, Kim SH, Choi SH, Sung H, Kim MN, Woo JH, et al. Clinical and radiological features of invasive pulmonary aspergillosis in transplant recipients and neutropenic patients. Transpl Infect Dis. 2010;12(4):309-15.

[112] de Heer K, Gerritsen MG, Visser CE, Leeflang MM. Galactomannan detection in broncho-alveolar lavage fluid for invasive aspergillosis in immunocompromised patients. Cochrane Database Syst Rev. 2019;5:CD012399.

[113] Leeflang MM, Debets-Ossenkopp YJ, Wang J, Visser CE, Scholten RJ, Hooft L, et al. Galactomannan detection for invasive aspergillosis in immunocompromised patients. Cochrane Database Syst Rev. 2015(12):CD007394.

[114] Chong GM, Maertens JA, Lagrou K, Driessen GJ, Cornelissen JJ, Rijnders BJ. Diagnostic Performance of Galactomannan Antigen Testing in Cerebrospinal Fluid. J Clin Microbiol. 2016;54(2):428-31.

[115] Duarte RF, Sanchez-Ortega I, Cuesta I, Arnan M, Patino B, Fernandez de Sevilla A, et al. Serum galactomannan-based early detection of invasive aspergillosis in hematology patients receiving effective antimold prophylaxis. Clin Infect Dis. 2014;59(12):1696-702.

[116] Cruciani M, White PL, Mengoli C, Loffler J, Morton CO, Klingspor L, et al. The impact of anti-mould prophylaxis on *Aspergillus* PCR blood testing for the diagnosis of invasive aspergillosis. J Antimicrob Chemother. 2021;76(3): 635-8.

[117] Mikulska M, Furfaro E, De Carolis E, Drago E, Pulzato I, Borghesi ML, et al. Use of *Aspergillus fumigatus* real-time PCR in bronchoalveolar lavage samples (BAL) for diagnosis of invasive aspergillosis, including azole-resistant cases, in high risk haematology patients: the need for a combined use with galactomannan. Med Mycol. 2019;57(8):987-96.

[118] Heldt S, Prattes J, Eigl S, Spiess B, Flick H, Rabensteiner J, et al. Diagnosis of invasive aspergillosis in hematological malignancy patients: Performance of cytokines, Asp LFD, and *Aspergillus* PCR in same day blood and bronchoalveolar lavage samples. J Infect. 2018;77(3):235-41.

[119] Shokouhi S, Mirzaei J, Sajadi MM, Javadi A. Comparison of serum PCR assay and histopathology for the diagnosis of invasive aspergillosis and mucormycosis in immunocompromised patients with sinus involvement. Curr Med Mycol. 2016;2(4):46-8.

[120] Dannaoui E, Gabriel F, Gaboyard M, Lagardere G, Audebert L, Quesne G, et al. Molecular Diagnosis of Invasive Aspergillosis and Detection of Azole Resistance by a Newly Commercialized PCR Kit. J Clin Microbiol. 2017;55(11):3210-8.

[121] White PL, Wingard JR, Bretagne S, Loffler J, Patterson TF, Slavin MA, et al. *Aspergillus* Polymerase Chain Reaction: Systematic Review of Evidence for Clinical Use in Comparison With Antigen Testing. Clin Infect Dis. 2015;61(8):1293-303.

[122] Freeman Weiss Z, Leon A, Koo S. The Evolving Landscape of Fungal

Diagnostics, Current and Emerging Microbiological Approaches. J Fungi (Basel). 2021;7(2).

[123] Loeffler J, Mengoli C, Springer J, Bretagne S, Cuenca-Estrella M, Klingspor L, et al. Analytical Comparison of In Vitro-Spiked Human Serum and Plasma for PCR-Based Detection of *Aspergillus fumigatus* DNA: a Study by the European *Aspergillus* PCR Initiative. J Clin Microbiol. 2015;53(9): 2838-45.

[124] White PL, Barnes RA, Springer J, Klingspor L, Cuenca-Estrella M, Morton CO, et al. Clinical Performance of *Aspergillus* PCR for Testing Serum and Plasma: a Study by the European *Aspergillus* PCR Initiative. J Clin Microbiol. 2015;53(9):2832-7.

[125] Blot SI, Taccone FS, Van den Abeele AM, Bulpa P, Meersseman W, Brusselaers N, et al. A clinical algorithm to diagnose invasive pulmonary aspergillosis in critically ill patients. Am J Respir Crit Care Med. 2012;186(1): 56-64.

[126] Verweij PE, Rijnders BJA, Bruggemann RJM, Azoulay E, Bassetti M, Blot S, et al. Review of influenza-associated pulmonary aspergillosis in ICU patients and proposal for a case definition: an expert opinion. Intensive Care Med. 2020;46(8):1524-35.

[127] Koehler P, Bassetti M, Chakrabarti A, Chen SCA, Colombo AL, Hoenigl M, et al. Defining and managing COVID-19-associated pulmonary aspergillosis: the 2020 ECMM/ISHAM consensus criteria for research and clinical guidance. The Lancet Infectious Diseases. 2020.

[128] Ullmann AJ, Aguado JM, Arikan-Akdagli S, Denning DW, Groll AH, Lagrou K, et al. Diagnosis and management of *Aspergillus* diseases: executive summary of the 2017

ESCMID-ECMM-ERS guideline. Clin Microbiol Infect. 2018;24 Suppl 1:e1-e38.

[129] Cuenca-Estrella M, Kett DH, Wauters J. Defining standards of CARE for invasive fungal diseases in the ICU. J Antimicrob Chemother. 2019;74(Suppl 2):ii9-ii15.

[130] Azoulay E, Afessa B. Diagnostic criteria for invasive pulmonary aspergillosis in critically ill patients. Am J Respir Crit Care Med. 2012;186(1): 8-10.

[131] Tissot F, Agrawal S, Pagano L, Petrikkos G, Groll AH, Skiada A, et al. ECIL-6 guidelines for the treatment of invasive candidiasis, aspergillosis and mucormycosis in leukemia and hematopoietic stem cell transplant patients. Haematologica. 2017;102(3): 433-44.

[132] Maertens JA, Girmenia C, Bruggemann RJ, Duarte RF, Kibbler CC, Ljungman P, et al. European guidelines for primary antifungal prophylaxis in adult haematology patients: summary of the updated recommendations from the European Conference on Infections in Leukaemia. J Antimicrob Chemother. 2018;73(12):3221-30.

[133] Wang J, Zhou M, Xu JY, Zhou RF, Chen B, Wan Y. Comparison of Antifungal Prophylaxis Drugs in Patients With Hematological Disease or Undergoing Hematopoietic Stem Cell Transplantation: A Systematic Review and Network Meta-analysis. JAMA Netw Open. 2020;3(10):e2017652.

[134] Garcia-Vidal C, Carratala J, Lortholary O. Defining standards of CARE for invasive fungal diseases in solid organ transplant patients. J Antimicrob Chemother. 2019;74(Suppl 2):ii16-ii20.

[135] Dolton MJ, McLachlan AJ. Voriconazole pharmacokinetics and exposure-response relationships:

assessing the links between exposure, efficacy and toxicity. Int J Antimicrob Agents. 2014;44(3):183-93.

[136] Mikus G, Scholz IM, Weiss J. Pharmacogenomics of the triazole antifungal agent voriconazole. Pharmacogenomics. 2011;12(6):861-72.

[137] Mitsani D, Nguyen MH, Shields RK, Toyoda Y, Kwak EJ, Silveira FP, et al. Prospective, observational study of voriconazole therapeutic drug monitoring among lung transplant recipients receiving prophylaxis: factors impacting levels of and associations between serum troughs, efficacy, and toxicity. Antimicrob Agents Chemother. 2012;56(5):2371-7.

[138] Elewa H, El-Mekaty E, El-Bardissy A, Ensom MH, Wilby KJ. Therapeutic Drug Monitoring of Voriconazole in the Management of Invasive Fungal Infections: A Critical Review. Clin Pharmacokinet. 2015;54(12):1223-35.

[139] Miceli MH, Kauffman CA. Isavuconazole: A New Broad-Spectrum Triazole Antifungal Agent. Clin Infect Dis. 2015;61(10):1558-65.

[140] Falci DR, Pasqualotto AC. Profile of isavuconazole and its potential in the treatment of severe invasive fungal infections. Infect Drug Resist. 2013;6:163-74.

[141] Livermore J, Hope W. Evaluation of the pharmacokinetics and clinical utility of isavuconazole for treatment of invasive fungal infections. Expert Opin Drug Metab Toxicol. 2012;8(6):759-65.

[142] Ellsworth M, Ostrosky-Zeichner L. Isavuconazole: Mechanism of Action, Clinical Efficacy, and Resistance. J Fungi (Basel). 2020;6(4).

[143] Keirns J, Desai A, Kowalski D, Lademacher C, Mujais S, Parker B, et al.

QT Interval Shortening With Isavuconazole: In Vitro and In Vivo Effects on Cardiac Repolarization. Clin Pharmacol Ther. 2017;101(6):782-90.

[144] Maertens JA, Raad, II, Marr KA, Patterson TF, Kontoyiannis DP, Cornely OA, et al. Isavuconazole versus voriconazole for primary treatment of invasive mould disease caused by *Aspergillus* and other filamentous fungi (SECURE): a phase 3, randomised-controlled, non-inferiority trial. Lancet. 2016;387(10020):760-9.

[145] Van Daele R, Spriet I, Maertens J. Posaconazole in prophylaxis and treatment of invasive fungal infections: a pharmacokinetic, pharmacodynamic and clinical evaluation. Expert Opin Drug Metab Toxicol. 2020;16(7):539-50.

[146] Courtney R, Wexler D, Radwanski E, Lim J, Laughlin M. Effect of food on the relative bioavailability of two oral formulations of posaconazole in healthy adults. Br J Clin Pharmacol. 2004;57(2):218-22.

[147] Anderson TM, Clay MC, Cioffi AG, Diaz KA, Hisao GS, Tuttle MD, et al. Amphotericin forms an extramembranous and fungicidal sterol sponge. Nat Chem Biol. 2014;10(5):400-6.

[148] Roden MM, Nelson LD, Knudsen TA, Jarosinski PF, Starling JM, Shiflett SE, et al. Triad of acute infusion-related reactions associated with liposomal amphotericin B: analysis of clinical and epidemiological characteristics. Clin Infect Dis. 2003;36(10):1213-20.

[149] Chen SC, Slavin MA, Sorrell TC. Echinocandin antifungal drugs in fungal infections: a comparison. Drugs. 2011;71(1):11-41.

[150] Denning DW. Echinocandin antifungal drugs. Lancet. 2003; 362(9390):1142-51.

[151] Wurthwein G, Cornely OA, Trame MN, Vehreschild JJ, Vehreschild MJ, Farowski F, et al. Population pharmacokinetics of escalating doses of caspofungin in a phase II study of patients with invasive aspergillosis. Antimicrob Agents Chemother. 2013;57(4):1664-71.

[152] Hiemenz JW, Raad, II, Maertens JA, Hachem RY, Saah AJ, Sable CA, et al. Efficacy of caspofungin as salvage therapy for invasive aspergillosis compared to standard therapy in a historical cohort. Eur J Clin Microbiol Infect Dis. 2010;29(11): 1387-94.

[153] Heinz WJ, Buchheidt D, Ullmann AJ. Clinical evidence for caspofungin monotherapy in the first-line and salvage therapy of invasive *Aspergillus* infections. Mycoses. 2016;59(8):480-93.

[154] Husain S, Camargo JF. Invasive Aspergillosis in solid-organ transplant recipients: Guidelines from the American Society of Transplantation Infectious Diseases Community of Practice. Clin Transplant. 2019;33(9): e13544.

[155] Husain S, Sole A, Alexander BD, Aslam S, Avery R, Benden C, et al. The 2015 International Society for Heart and Lung Transplantation Guidelines for the management of fungal infections in mechanical circulatory support and cardiothoracic organ transplant recipients: Executive summary. J Heart Lung Transplant. 2016;35(3):261-82.

[156] Jeanvoine A, Rocchi S, Bellanger AP, Reboux G, Millon L. Azole-resistant *Aspergillus fumigatus*: A global phenomenon originating in the environment? Med Mal Infect. 2020;50(5):389-95.

[157] Berkow EL, Nunnally NS, Bandea A, Kuykendall R, Beer K, Lockhart SR. Detection of TR34/L98H

CYP51A Mutation through Passive Surveillance for Azole-Resistant *Aspergillus fumigatus* in the United States from 2015 to 2017. Antimicrob Agents Chemother. 2018;62(5).

[158] Chowdhary A, Kathuria S, Xu J, Meis JF. Emergence of azole-resistant *aspergillus fumigatus* strains due to agricultural azole use creates an increasing threat to human health. PLoS Pathog. 2013;9(10):e1003633.

[159] Toyotome T. Resistance in the Environmental Pathogenic Fungus *Aspergillus fumigatus*. Med Mycol J. 2019;60(3):61-3.

[160] Resendiz-Sharpe A, Dewaele K, Merckx R, Bustamante B, Vega-Gomez MC, Rolon M, et al. Triazole-Resistance in Environmental *Aspergillus fumigatus* in Latin American and African Countries. J Fungi (Basel). 2021;7(4).

[161] Toda M, Beer KD, Kuivila KM, Chiller TM, Jackson BR. Trends in Agricultural Triazole Fungicide Use in the United States, 1992-2016 and Possible Implications for Antifungal-Resistant Fungi in Human Disease. Environ Health Perspect. 2021;129(5): 55001.

[162] van der Linden JW, Snelders E, Kampinga GA, Rijnders BJ, Mattsson E, Debets-Ossenkopp YJ, et al. Clinical implications of azole resistance in *Aspergillus fumigatus*, The Netherlands, 2007-2009. Emerg Infect Dis. 2011;17(10):1846-54.

[163] Chowdhary A, Kathuria S, Randhawa HS, Gaur SN, Klaassen CH, Meis JF. Isolation of multiple-triazole-resistant *Aspergillus fumigatus* strains carrying the TR/L98H mutations in the cyp51A gene in India. J Antimicrob Chemother. 2012;67(2):362-6.

[164] Snelders E, van der Lee HA, Kuijpers J, Rijs AJ, Varga J, Samson RA,

et al. Emergence of azole resistance in *Aspergillus fumigatus* and spread of a single resistance mechanism. PLoS Med. 2008;5(11):e219.

[165] Meis JF, Chowdhary A, Rhodes JL, Fisher MC, Verweij PE. Clinical implications of globally emerging azole resistance in *Aspergillus fumigatus*. Philos Trans R Soc Lond B Biol Sci. 2016;371(1709).

[166] Mellado E, Garcia-Effron G, Alcazar-Fuoli L, Melchers WJ, Verweij PE, Cuenca-Estrella M, et al. A new *Aspergillus fumigatus* resistance mechanism conferring in vitro cross-resistance to azole antifungals involves a combination of cyp51A alterations. Antimicrob Agents Chemother. 2007;51(6):1897-904.

[167] Snelders E, Huis In 't Veld RA, Rijs AJ, Kema GH, Melchers WJ, Verweij PE. Possible environmental origin of resistance of *Aspergillus fumigatus* to medical triazoles. Applied and environmental microbiology. 2009;75(12):4053-7.

[168] Mortensen KL, Mellado E, Lass-Florl C, Rodriguez-Tudela JL, Johansen HK, Arendrup MC. Environmental study of azole-resistant *Aspergillus fumigatus* and other aspergilli in Austria, Denmark, and Spain. Antimicrob Agents Chemother. 2010;54(11):4545-9.

[169] Chowdhary A, Kathuria S, Xu J, Sharma C, Sundar G, Singh PK, et al. Clonal expansion and emergence of environmental multiple-triazole-resistant *Aspergillus fumigatus* strains carrying the TR(3)(4)/L98H mutations in the cyp51A gene in India. PLoS One. 2012;7(12):e52871.

[170] Badali H, Vaezi A, Haghani I, Yazdanparast SA, Hedayati MT, Mousavi B, et al. Environmental study of azole-resistant *Aspergillus fumigatus* with TR34/L98H mutations in the cyp51A gene in Iran. Mycoses. 2013;56(6):659-63.

[171] van der Linden JW, Camps SM, Kampinga GA, Arends JP, Debets-Ossenkopp YJ, Haas PJ, et al. Aspergillosis due to voriconazole highly resistant *Aspergillus fumigatus* and recovery of genetically related resistant isolates from domiciles. Clin Infect Dis. 2013;57(4):513-20.

[172] Snelders E, Camps SM, Karawajczyk A, Schaftenaar G, Kema GH, van der Lee HA, et al. Triazole fungicides can induce cross-resistance to medical triazoles in *Aspergillus fumigatus*. PLoS One. 2012;7(3):e31801.

[173] Jorgensen KM, Helleberg M, Hare RK, Jorgensen LN, Arendrup MC. Dissection of the Activity of Agricultural Fungicides against Clinical *Aspergillus* Isolates with and without Environmentally and Medically Induced Azole Resistance. J Fungi (Basel). 2021;7(3).

[174] Bassetti M, Vena A, Bouza E, Peghin M, Munoz P, Righi E, et al. Antifungal susceptibility testing in *Candida*, *Aspergillus* and *Cryptococcus* infections: are the MICs useful for clinicians? Clin Microbiol Infect. 2020;26(8):1024-33.

[175] CLSI. Performance Standards for Antifungal Susceptibility Testing of Filamentous Fungi. 2nd ed. CLSI supplement M61. Wayne, PA: Clinical and Laboratory Standards Institute; 2020.

[176] The European Committee on Antimicrobial Susceptibility Testing. Overview of antifungal ECOFFs and clinical breakpoints for yeasts, moulds and dermatophytes using the EUCAST E.Def 7.3, E.Def 9.3 and E.Def 11.0 procedures. Version 2, 2020. http://www.eucast.org.

[177] CLSI. Epidemiological Cutoff Values for Antifungal Susceptibility Testing. 3rd ed. CLSI supplement M59. Wayne, PA: Clinical and Laboratory Standards Institute; 2020.

[178] Wiederhold NP, Patterson TF. Emergence of Azole Resistance in *Aspergillus*. Semin Respir Crit Care Med. 2015;36(5):673-80.

[179] Garcia-Effron G. Molecular Markers of Antifungal Resistance: Potential Uses in Routine Practice and Future Perspectives. J Fungi (Basel). 2021;7(3).

[180] Verweij PE, Ananda-Rajah M, Andes D, Arendrup MC, Bruggemann RJ, Chowdhary A, et al. International expert opinion on the management of infection caused by azole-resistant *Aspergillus fumigatus*. Drug Resist Updat. 2015;21-22:30-40.

[181] Scorzoni L, Sangalli-Leite F, de Lacorte Singulani J, de Paula ESAC, Costa-Orlandi CB, Fusco-Almeida AM, et al. Searching new antifungals: The use of in vitro and in vivo methods for evaluation of natural compounds. J Microbiol Methods. 2016;123:68-78.

[182] Perfect JR. The antifungal pipeline: a reality check. Nat Rev Drug Discov. 2017;16(9):603-16.

[183] CLSI. Reference Method for Broth Dilution Antifungal Susceptibility Testing of Filamentous Fungi. 3ed ed. CLSI standard M38. Wayne, PA: Clinical and Laboratory Standards Institute; 2017.

[184] Odds FC. Synergy, antagonism, and what the chequerboard puts between them. J Antimicrob Chemother. 2003;52(1):1.

[185] Meletiadis J, Pournaras S, Roilides E, Walsh TJ. Defining fractional inhibitory concentration index cutoffs for additive interactions based on

self-drug additive combinations, Monte Carlo simulation analysis, and in vitro-in vivo correlation data for antifungal drug combinations against *Aspergillus fumigatus*. Antimicrob Agents Chemother. 2010;54(2):602-9.

[186] Plabutong N, Ekronarongchai S, Niwetboworchai N, Edwards SW, Virakul S, Chiewchengchol D, et al. The Inhibitory Effect of Validamycin A on *Aspergillus flavus*. Int J Microbiol. 2020;2020:3972415.

[187] Umscheid CA, Margolis DJ, Grossman CE. Key concepts of clinical trials: a narrative review. Postgrad Med. 2011;123(5):194-204.

[188] Aigner M, Lass-Florl C. Encochleated Amphotericin B: Is the Oral Availability of Amphotericin B Finally Reached? J Fungi (Basel). 2020;6(2).

[189] Santangelo R, Paderu P, Delmas G, Chen ZW, Mannino R, Zarif L, et al. Efficacy of oral cochleate-amphotericin B in a mouse model of systemic candidiasis. Antimicrob Agents Chemother. 2000;44(9):2356-60.

[190] Zarif L, Graybill JR, Perlin D, Najvar L, Bocanegra R, Mannino RJ. Antifungal activity of amphotericin B cochleates against *Candida albicans* infection in a mouse model. Antimicrob Agents Chemother. 2000;44(6):1463-9.

[191] Lu R, Hollingsworth C, Qiu J, Wang A, Hughes E, Xin X, et al. Efficacy of Oral Encochleated Amphotericin B in a Mouse Model of Cryptococcal Meningoencephalitis. mBio. 2019;10(3).

[192] Skipper CP, Atukunda M, Stadelman A, Engen NW, Bangdiwala AS, Hullsiek KH, et al. Phase I EnACT Trial of the Safety and Tolerability of a Novel Oral Formulation of Amphotericin B. Antimicrob Agents Chemother. 2020;64(10).

[193] Kovacs R, Toth Z, Locke JB, Forgacs L, Kardos G, Nagy F, et al. Comparison of In Vitro Killing Activity of Rezafungin, Anidulafungin, Caspofungin, and Micafungin against Four *Candida auris* Clades in RPMI-1640 in the Absence and Presence of Human Serum. Microorganisms. 2021;9(4).

[194] Ham YY, Lewis JS, 2nd, Thompson GR, 3rd. Rezafungin: a novel antifungal for the treatment of invasive candidiasis. Future Microbiol. 2021;16:27-36.

[195] Miesel L, Cushion MT, Ashbaugh A, Lopez SR, Ong V. Efficacy of Rezafungin in Prophylactic Mouse Models of Invasive Candidiasis, Aspergillosis, and Pneumocystis Pneumonia. Antimicrob Agents Chemother. 2021;65(3).

[196] Garcia-Effron G. Rezafungin-Mechanisms of Action, Susceptibility and Resistance: Similarities and Differences with the Other Echinocandins. J Fungi (Basel). 2020;6(4).

[197] Zhao Y, Perlin DS. Review of the Novel Echinocandin Antifungal Rezafungin: Animal Studies and Clinical Data. J Fungi (Basel). 2020;6(4).

[198] Wiederhold NP, Najvar LK, Jaramillo R, Olivo M, Wickes BL, Catano G, et al. Extended-Interval Dosing of Rezafungin against Azole-Resistant *Aspergillus fumigatus*. Antimicrob Agents Chemother. 2019;63(10).

[199] Wiederhold NP, Locke JB, Daruwala P, Bartizal K. Rezafungin (CD101) demonstrates potent in vitro activity against *Aspergillus*, including azole-resistant *Aspergillus fumigatus* isolates and cryptic species. J Antimicrob Chemother. 2018;73(11):3063-7.

[200] Sofjan AK, Mitchell A, Shah DN, Nguyen T, Sim M, Trojcak A, et al. Rezafungin (CD101), a next-generation echinocandin: A systematic literature review and assessment of possible place in therapy. J Glob Antimicrob Resist. 2018;14:58-64.

[201] Helleberg M, Jorgensen KM, Hare RK, Datcu R, Chowdhary A, Arendrup MC. Rezafungin In Vitro Activity against Contemporary Nordic Clinical *Candida* Isolates and *Candida auris* Determined by the EUCAST Reference Method. Antimicrob Agents Chemother. 2020;64(4).

[202] Pfaller MA, Carvalhaes C, Messer SA, Rhomberg PR, Castanheira M. Activity of a Long-Acting Echinocandin, Rezafungin, and Comparator Antifungal Agents Tested against Contemporary Invasive Fungal Isolates (SENTRY Program, 2016 to 2018). Antimicrob Agents Chemother. 2020;64(4).

[203] Thompson GR, Soriano A, Skoutelis A, Vazquez JA, Honore PM, Horcajada JP, et al. Rezafungin versus Caspofungin in a Phase 2, Randomized, Double-Blind Study for the Treatment of Candidemia and Invasive Candidiasis- The STRIVE Trial. Clin Infect Dis. 2020.

[204] Jallow S, Govender NP. Ibrexafungerp: A First-in-Class Oral Triterpenoid Glucan Synthase Inhibitor. J Fungi (Basel). 2021;7(3).

[205] Apgar JM, Wilkening RR, Parker DL, Jr., Meng D, Wildonger KJ, Sperbeck D, et al. Ibrexafungerp: An orally active beta-1,3-glucan synthesis inhibitor. Bioorg Med Chem Lett. 2021;32:127661.

[206] Ghannoum M, Arendrup MC, Chaturvedi VP, Lockhart SR, McCormick TS, Chaturvedi S, et al. Ibrexafungerp: A Novel Oral Triterpenoid Antifungal in

Development for the Treatment of *Candida auris* Infections. Antibiotics (Basel). 2020;9(9).

[207] Petraitis V, Petraitiene R, Katragkou A, Maung BBW, Naing E, Kavaliauskas P, et al. Combination Therapy with Ibrexafungerp (Formerly SCY-078), a First-in-Class Triterpenoid Inhibitor of (1-->3)-beta-d-Glucan Synthesis, and Isavuconazole for Treatment of Experimental Invasive Pulmonary Aspergillosis. Antimicrob Agents Chemother. 2020;64(6).

[208] Davis MR, Donnelley MA, Thompson GR. Ibrexafungerp: A novel oral glucan synthase inhibitor. Med Mycol. 2020;58(5):579-92.

[209] Spec A, Pullman J, Thompson GR, Powderly WG, Tobin EH, Vazquez J, et al. MSG-10: a Phase 2 study of oral ibrexafungerp (SCY-078) following initial echinocandin therapy in non-neutropenic patients with invasive candidiasis. J Antimicrob Chemother. 2019;74(10):3056-62.

[210] Larkin EL, Long L, Isham N, Borroto-Esoda K, Barat S, Angulo D, et al. A Novel 1,3-Beta-d-Glucan Inhibitor, Ibrexafungerp (Formerly SCY-078), Shows Potent Activity in the Lower pH Environment of Vulvovaginitis. Antimicrob Agents Chemother. 2019;63(5).

[211] Garvey EP, Sharp AD, Warn PA, Yates CM, Atari M, Thomas S, et al. The novel fungal CYP51 inhibitor VT-1598 displays classic dose-dependent antifungal activity in murine models of invasive aspergillosis. Med Mycol. 2020;58(4):505-13.

[212] Hargrove TY, Garvey EP, Hoekstra WJ, Yates CM, Wawrzak Z, Rachakonda G, et al. Crystal Structure of the New Investigational Drug Candidate VT-1598 in Complex with *Aspergillus fumigatus* Sterol 14alpha-Demethylase Provides Insights into Its

Broad-Spectrum Antifungal Activity. Antimicrob Agents Chemother. 2017;61(7).

[213] Elewski B, Brand S, Degenhardt T, Curelop S, Pollak R, Schotzinger R, et al. A phase II, randomized, double-blind, placebo-controlled, dose-ranging study to evaluate the efficacy and safety of VT-1161 oral tablets in the treatment of patients with distal and lateral subungual onychomycosis of the toenail. Br J Dermatol. 2021;184(2): 270-80.

[214] Monk BC, Keniya MV, Sabherwal M, Wilson RK, Graham DO, Hassan HF, et al. Azole Resistance Reduces Susceptibility to the Tetrazole Antifungal VT-1161. Antimicrob Agents Chemother. 2019;63(1).

[215] Wiederhold NP, Xu X, Wang A, Najvar LK, Garvey EP, Ottinger EA, et al. In Vivo Efficacy of VT-1129 against Experimental Cryptococcal Meningitis with the Use of a Loading Dose-Maintenance Dose Administration Strategy. Antimicrob Agents Chemother. 2018;62(11).

[216] Wiederhold NP, Najvar LK, Garvey EP, Brand SR, Xu X, Ottinger EA, et al. The Fungal Cyp51 Inhibitor VT-1129 Is Efficacious in an Experimental Model of Cryptococcal Meningitis. Antimicrob Agents Chemother. 2018;62(9).

[217] Schell WA, Jones AM, Garvey EP, Hoekstra WJ, Schotzinger RJ, Alexander BD. Fungal CYP51 Inhibitors VT-1161 and VT-1129 Exhibit Strong In Vitro Activity against *Candida glabrata* and *C. krusei* Isolates Clinically Resistant to Azole and Echinocandin Antifungal Compounds. Antimicrob Agents Chemother. 2017;61(3).

[218] Nielsen K, Vedula P, Smith KD, Meya DB, Garvey EP, Hoekstra WJ, et al. Activity of VT-1129 against *Cryptococcus neoformans* clinical isolates with high

fluconazole MICs. Med Mycol. 2017;55(4):453-6.

[219] Warrilow AG, Parker JE, Price CL, Nes WD, Garvey EP, Hoekstra WJ, et al. The Investigational Drug VT-1129 Is a Highly Potent Inhibitor of *Cryptococcus* Species CYP51 but Only Weakly Inhibits the Human Enzyme. Antimicrob Agents Chemother. 2016;60(8):4530-8.

[220] Lockhart SR, Fothergill AW, Iqbal N, Bolden CB, Grossman NT, Garvey EP, et al. The Investigational Fungal Cyp51 Inhibitor VT-1129 Demonstrates Potent In Vitro Activity against *Cryptococcus neoformans* and *Cryptococcus gattii*. Antimicrob Agents Chemother. 2016;60(4):2528-31.

[221] Murray A, Cass L, Ito K, Pagani N, Armstrong-James D, Dalal P, et al. PC945, a Novel Inhaled Antifungal Agent, for the Treatment of Respiratory Fungal Infections. J Fungi (Basel). 2020;6(4).

[222] Cass L, Murray A, Davis A, Woodward K, Albayaty M, Ito K, et al. Safety and nonclinical and clinical pharmacokinetics of PC945, a novel inhaled triazole antifungal agent. Pharmacol Res Perspect. 2021;9(1): e00690.

[223] Pagani N, Armstrong-James D, Reed A. Successful salvage therapy for fungal bronchial anastomotic infection after -lung transplantation with an inhaled triazole anti-fungal PC945. J Heart Lung Transplant. 2020;39(12): 1505-6.

[224] Rudramurthy SM, Colley T, Abdolrasouli A, Ashman J, Dhaliwal M, Kaur H, et al. In vitro antifungal activity of a novel topical triazole PC945 against emerging yeast *Candida auris*. J Antimicrob Chemother. 2019;74(10):2943-9.

[225] Colley T, Sehra G, Daly L, Kimura G, Nakaoki T, Nishimoto Y, et al.

Antifungal synergy of a topical triazole, PC945, with a systemic triazole against respiratory *Aspergillus fumigatus* infection. Scientific reports. 2019;9(1):9482.

[226] Kimura G, Nakaoki T, Colley T, Rapeport G, Strong P, Ito K, et al. In Vivo Biomarker Analysis of the Effects of Intranasally Dosed PC945, a Novel Antifungal Triazole, on *Aspergillus fumigatus* Infection in Immunocompromised Mice. Antimicrob Agents Chemother. 2017;61(9).

[227] Colley T, Alanio A, Kelly SL, Sehra G, Kizawa Y, Warrilow AGS, et al. In Vitro and In Vivo Antifungal Profile of a Novel and Long-Acting Inhaled Azole, PC945, on *Aspergillus fumigatus* Infection. Antimicrob Agents Chemother. 2017;61(5).

[228] Badali H, Patterson HP, Sanders CJ, Mermella B, Gibas CFC, Ibrahim AS, et al. Manogepix, the Active Moiety of the Investigational Agent Fosmanogepix, Demonstrates In vitro Activity against Members of the *Fusarium oxysporum* and *Fusarium solani* Species Complexes. Antimicrob Agents Chemother. 2021.

[229] Lee A, Wang N, Carter CL, Zimmerman M, Dartois V, Shaw KJ, et al. Therapeutic Potential of Fosmanogepix (APX001) for Intra-abdominal Candidiasis: from Lesion Penetration to Efficacy in a Mouse Model. Antimicrob Agents Chemother. 2021;65(4).

[230] Petraitiene R, Petraitis V, Maung BBW, Mansbach RS, Hodges MR, Finkelman MA, et al. Efficacy and Pharmacokinetics of Fosmanogepix (APX001) in the Treatment of *Candida* Endophthalmitis and Hematogenous Meningoencephalitis in Nonneutropenic Rabbits. Antimicrob Agents Chemother. 2021;65(3).

[231] Shaw KJ, Ibrahim AS. Fosmanogepix: A Review of the First-in-Class Broad Spectrum Agent for the Treatment of Invasive Fungal Infections. J Fungi (Basel). 2020;6(4).

[232] Gebremariam T, Alkhazraji S, Alqarihi A, Wiederhold NP, Shaw KJ, Patterson TF, et al. Fosmanogepix (APX001) Is Effective in the Treatment of Pulmonary Murine Mucormycosis Due to Rhizopus arrhizus. Antimicrob Agents Chemother. 2020;64(6).

[233] Alkhazraji S, Gebremariam T, Alqarihi A, Gu Y, Mamouei Z, Singh S, et al. Fosmanogepix (APX001) Is Effective in the Treatment of Immunocompromised Mice Infected with Invasive Pulmonary Scedosporiosis or Disseminated Fusariosis. Antimicrob Agents Chemother. 2020;64(3).

[234] Gebremariam T, Alkhazraji S, Gu Y, Singh S, Alqarihi A, Shaw KJ, et al. Galactomannan Is a Biomarker of Fosmanogepix (APX001) Efficacy in Treating Experimental Invasive Pulmonary Aspergillosis. Antimicrob Agents Chemother. 2019;64(1).

[235] Wiederhold NP, Najvar LK, Shaw KJ, Jaramillo R, Patterson H, Olivo M, et al. Efficacy of Delayed Therapy with Fosmanogepix (APX001) in a Murine Model of Candida auris Invasive Candidiasis. Antimicrob Agents Chemother. 2019;63(11).

[236] Shaw KJ, Schell WA, Covel J, Duboc G, Giamberardino C, Kapoor M, et al. In Vitro and In Vivo Evaluation of APX001A/APX001 and Other Gwt1 Inhibitors against Cryptococcus. Antimicrob Agents Chemother. 2018;62(8).

[237] Su H, Zhu M, Tsui CK, van der Lee H, Tehupeiory-Kooreman M, Zoll J, et al. Potency of olorofim (F901318) compared to contemporary antifungal agents against clinical Aspergillus fumigatus isolates, and review of azole resistance phenotype and genotype epidemiology in China. Antimicrob Agents Chemother. 2021.

[238] Singh A, Singh P, Meis JF, Chowdhary A. In vitro activity of the novel antifungal olorofim against dermatophytes and opportunistic moulds including Penicillium and Talaromyces species. J Antimicrob Chemother. 2021;76(5):1229-33.

[239] du Pre S, Beckmann N, Almeida MC, Sibley GEM, Law D, Brand AC, et al. Effect of the Novel Antifungal Drug F901318 (Olorofim) on Growth and Viability of Aspergillus fumigatus. Antimicrob Agents Chemother. 2018;62(8).

[240] Wiederhold NP. Review of T-2307, an Investigational Agent That Causes Collapse of Fungal Mitochondrial Membrane Potential. J Fungi (Basel). 2021;7(2).

[241] Yamashita K, Miyazaki T, Fukuda Y, Mitsuyama J, Saijo T, Shimamura S, et al. The Novel Arylamidine T-2307 Selectively Disrupts Yeast Mitochondrial Function by Inhibiting Respiratory Chain Complexes. Antimicrob Agents Chemother. 2019;63(8).

[242] Mitsuyama J, Nomura N, Hashimoto K, Yamada E, Nishikawa H, Kaeriyama M, et al. In vitro and in vivo antifungal activities of T-2307, a novel arylamidine. Antimicrob Agents Chemother. 2008;52(4):1318-24.

[243] Mammen MP, Armas D, Hughes FH, Hopkins AM, Fisher CL, Resch PA, et al. First-in-Human Phase 1 Study To Assess Safety, Tolerability, and Pharmacokinetics of a Novel Antifungal Drug, VL-2397, in Healthy Adults. Antimicrob Agents Chemother. 2019;63(11).

[244] Dietl AM, Misslinger M, Aguiar MM, Ivashov V, Teis D, Pfister J,

et al. The Siderophore Transporter Sit1 Determines Susceptibility to the Antifungal VL-2397. Antimicrob Agents Chemother. 2019;63(10).

[245] Kovanda LL, Sullivan SM, Smith LR, Desai AV, Bonate PL, Hope WW. Population Pharmacokinetic Modeling of VL-2397, a Novel Systemic Antifungal Agent: Analysis of a Single- and Multiple-Ascending-Dose Study in Healthy Subjects. Antimicrob Agents Chemother. 2019;63(6).

[246] Mahmoudi S, Rezaie S, Daie Ghazvini R, Hashemi SJ, Badali H, Foroumadi A, et al. In Vitro Interaction of Geldanamycin with Triazoles and Echinocandins Against Common and Emerging *Candida* Species. Mycopathologia. 2019;184(5):607-13.

[247] Ma C, Chen J, Li P. Geldanamycin induces apoptosis and inhibits inflammation in fibroblast-like synoviocytes isolated from rheumatoid arthritis patients. J Cell Biochem. 2019;120(9):16254-63.

[248] Ochel HJ, Eichhorn K, Gademann G. Geldanamycin: the prototype of a class of antitumor drugs targeting the heat shock protein 90 family of molecular chaperones. Cell Stress Chaperones. 2001;6(2):105-12.

[249] High KP. The antimicrobial activities of cyclosporine, FK506, and rapamycin. Transplantation. 1994;57(12):1689-700.

[250] Gao L, Sun Y. In vitro interactions of antifungal agents and tacrolimus against *Aspergillus* biofilms. Antimicrob Agents Chemother. 2015;59(11):7097-9.

[251] Lee Y, Lee KT, Lee SJ, Beom JY, Hwangbo A, Jung JA, et al. In Vitro and In Vivo Assessment of FK506 Analogs as Novel Antifungal Drug Candidates. Antimicrob Agents Chemother. 2018;62(11).

[252] Pandit RT. Antifungal effects of cyclosporine A. Cornea. 2003;22(1):92-3.

[253] Robbins N, Leach MD, Cowen LE. Lysine deacetylases Hda1 and Rpd3 regulate Hsp90 function thereby governing fungal drug resistance. Cell Rep. 2012;2(4):878-88.

[254] Wurtele H, Tsao S, Lepine G, Mullick A, Tremblay J, Drogaris P, et al. Modulation of histone H3 lysine 56 acetylation as an antifungal therapeutic strategy. Nat Med. 2010;16(7):774-80.

[255] Zhong W, Jeffries MW, Georgopapadakou NH. Inhibition of inositol phosphorylceramide synthase by aureobasidin A in *Candida* and *Aspergillus* species. Antimicrob Agents Chemother. 2000;44(3):651-3.

[256] Teymuri M, Shams-Ghahfarokhi M, Razzaghi-Abyaneh M. Inhibitory effects and mechanism of antifungal action of the natural cyclic depsipeptide, aureobasidin A against *Cryptococcus neoformans*. Bioorg Med Chem Lett. 2021;41:128013.

[257] Munusamy K, Vadivelu J, Tay ST. A study on *Candida* biofilm growth characteristics and its susceptibility to aureobasidin A. Rev Iberoam Micol. 2018;35(2):68-72.

[258] Tan HW, Tay ST. The inhibitory effects of aureobasidin A on *Candida* planktonic and biofilm cells. Mycoses. 2013;56(2):150-6.

[259] Larwood DJ. Nikkomycin Z-Ready to Meet the Promise? J Fungi (Basel). 2020;6(4).

[260] Lazzarini C, Haranahalli K, Rieger R, Ananthula HK, Desai PB, Ashbaugh A, et al. Acylhydrazones as Antifungal Agents Targeting the Synthesis of Fungal Sphingolipids. Antimicrob Agents Chemother. 2018;62(5).

[261] Lazzarini C, Haranahalli K, McCarthy JB, Mallamo J, Ojima I, Del Poeta M. Preclinical Evaluation of Acylhydrazone SB-AF-1002 as a Novel Broad-Spectrum Antifungal Agent. Antimicrob Agents Chemother. 2020;64(9).

[262] Yu Y, Albrecht K, Groll J, Beilhack A. Innovative therapies for invasive fungal infections in preclinical and clinical development. Expert Opin Investig Drugs. 2020;29(9):961-71.

[263] Svanstrom A, van Leeuwen MR, Dijksterhuis J, Melin P. Trehalose synthesis in *Aspergillus niger*: characterization of six homologous genes, all with conserved orthologs in related species. BMC Microbiol. 2014;14:90.

[264] Al-Bader N, Vanier G, Liu H, Gravelat FN, Urb M, Hoareau CM, et al. Role of trehalose biosynthesis in *Aspergillus fumigatus* development, stress response, and virulence. Infect Immun. 2010;78(7):3007-18.

[265] Puttikamonkul S, Willger SD, Grahl N, Perfect JR, Movahed N, Bothner B, et al. Trehalose 6-phosphate phosphatase is required for cell wall integrity and fungal virulence but not trehalose biosynthesis in the human fungal pathogen *Aspergillus fumigatus*. Mol Microbiol. 2010;77(4):891-911.

[266] Shibata M, Mori K, Hamashima M. Inhibition of hyphal extension factor formation by validamycin in *Rhizoctonia solani*. J Antibiot (Tokyo). 1982;35(10): 1422-3.

[267] Shibata M, Uyeda M, Mori K. Stimulation of the extension of validamycin-inhibited hyphae by the hyphal extension factor present in *Rhizoctonia solani*. J Antibiot (Tokyo). 1981;34(4):447-51.

[268] Shibata M, Uyeda M, Mori K. Reversal of validamycin inhibition by the hyphal extract of *Rhizoctonia solani*. J Antibiot (Tokyo). 1980;33(6):679-81.

[269] Suami T, Ogawa S, Chida N. The revised structure of validamycin A. J Antibiot (Tokyo). 1980;33(1):98-9.

[270] Guirao-Abad JP, Sanchez-Fresneda R, Valentin E, Martinez-Esparza M, Arguelles JC. Analysis of validamycin as a potential antifungal compound against *Candida albicans*. Int Microbiol. 2013;16(4):217-25.

[271] Ji Y, Yang F, Ma D, Zhang J, Wan Z, Liu W, et al. HOG-MAPK signaling regulates the adaptive responses of *Aspergillus fumigatus* to thermal stress and other related stress. Mycopathologia. 2012;174(4):273-82.

[272] Day AM, Quinn J. Stress-Activated Protein Kinases in Human Fungal Pathogens. Front Cell Infect Microbiol. 2019;9:261.

[273] Hagiwara D, Takahashi H, Kusuya Y, Kawamoto S, Kamei K, Gonoi T. Comparative transcriptome analysis revealing dormant conidia and germination associated genes in *Aspergillus* species: an essential role for AtfA in conidial dormancy. BMC genomics. 2016;17:358.

[274] Hagiwara D, Suzuki S, Kamei K, Gonoi T, Kawamoto S. The role of AtfA and HOG MAPK pathway in stress tolerance in conidia of *Aspergillus fumigatus*. Fungal Genet Biol. 2014;73:138-49.

Chapter 3

Immunopathogenesis of Aspergillosis

Shreya Singh, Rimjhim Kanaujia
and Shivaprakash M. Rudramurthy

Abstract

Aspergillus species are ubiquitous saprophytes and opportunistic pathogens causing wide spectrum of diseases in humans depending on the host immune status. Following pathogen entry, various soluble bronchopulmonary factors enhance conidial clearance. However, due to virulence factors and poor host immune response *Aspergillus* conidia bind and damage the airway epithelium. The host immune cells like neutrophils and macrophages recognise *Aspergillus* spp. through various pathogen recognition receptors and form reactive oxygen species which mediate conidial killing. Neutrophils also attack extracellular hyphae by oxidative attack, non-oxidative granule proteins and neutrophil extracellular traps. In case of adaptive immunity, Th1 cells are crucial sources of IFN-γ mediated protective immunity. The Th17 also display a highly pro-inflammatory which is counterbalanced by a Treg cell. B cells and antibodies also enhance fungal clearance although excessive IgE production may result in atopy. The immune responses are influenced by changes in production of short-chain fatty acids by the gut microbiome which primes cells toward Th2 responses, and this is synchronized by the Innate lymphoid cells. This review provides comprehensive knowledge of various virulence factors of *Aspergillus*, antifungal host defences including innate and humoral immune response and regulation of host immunity by microbiome.

Keywords: Immunity, pathogenesis, aspergillus, genetic polymorphism, virulence

1. Introduction

Aspergillus species are globally ubiquitous saprophytes and are also opportunistic pathogens which have evolved in the environment and adapted to invade and proliferate within the human host. It can cause serious invasive infections. Invasive aspergillosis (IA) is associated with high mortality and morbidity which makes it essential to understand the factors involved in disease pathogenesis. The interplay between *Aspergillus* spp. and various components of the host immune system influences disease progression. Agent factors such as conidia size, temperature tolerance, hydrophobin /melanin expression etc. which contribute to virulence must be studied. Additionally, comprehensive knowledge of the host defenses, innate and humoral immune response, genetic susceptibility to *Aspergillus* and the role of microbiome in modulating immune response is important to study the disease immunopathogenesis.

In the genus *Aspergillus, Aspergillus fumigatus* is most commonly reported from human infections, followed by *A. flavus*, *A. terreus* and other uncommon species like *A. niger* and *A. nidulans* [1, 2]. It can cause plethora of infections, depending on the immune status of the host as immunocompetent individuals with asthma or cystic fibrosis are predisposed to a hypersensitive response while Invasive aspergillosis (IA) is seen in severely immunocompromised patients.

A better understanding of the interplay between the host immune system and *Aspergillus* is important to understand disease pathology and can provide us with useful insights regarding potential therapeutic targets. In this review, we will thus discuss the pathogen related virulence factors, clinical spectrum of diseases caused by it, its interaction with various components of the host immune system, factors involved in regulating the anti-fungal immune response and will also give an overview of the genetic polymorphisms in immune pathways that predispose to aspergillosis. *Aspergillus* and disease pathology and progression are the result of both fungal growth and the host response.

2. Virulence factors

The various virulence factors involved in the pathogenesis of aspergillosis are summarized in **Table 1.**

	Function	Gene(s) involved	Reference
Enzymes			
Superoxide dismutases (SODs)	Oxidative stress defense	*SOD* genes	[3]
Protease	Degradation of host structural barriers		
1. Serine protease	Degrades elastin.	36-kDa	[4]
2. Metalloproteinase	Degrades fibrinogen and laminin.	23-kDa	[5]
3. Aspartic (acid) proteinase	Assist in host cell invasion of the hyphae.		[6, 7]
Catalase	ROS scavengers. Breakdown hydrogen peroxide (H_2O_2) to oxygen and water.	*catA* - conidium-specific gene *cat2* - mycelium-specific gene	[8, 9]
Toxins			
1. Gliotoxin	Inhibits macrophage phagocytosis.	18-kDa cytotoxin	[10, 11]
2. Restrictocin	Induces fragmentation and apoptosis of DNA in macrophages. Inhibition of T-cell activation.	gene cluster of aflatoxin	[12, 13]
3. Aflatoxin	RNA nuclease activity by cleavage of the phosphodiester bond in the 28S rRNA of eukaryotic ribosomes	biosynthesis regulated by *AflCDC14*	[14]
	Induces DNA adducts causing genetic changes in cells responsible for carcinogenic potential *in vitro*. Also, epidemiologically to hepatocellular carcinoma.		
Others			
1. Melanin	Masking of beta (1,3)-glucan. Delay macrophage activation. ROS scavengers.	*pksP* - polyketide synthase gene	[15, 16]
2. Rodlets	Rodlet proteins form hydrophobic layer around *Aspergillus* conidia and helps in its dispersal. ROS scavengers.	*rodA* gene	[17]

Table 1.
Virulence factors of Aspergillus *species.*

3. Risk factors and clinical spectrum

An elaborate range of diseases can be caused by *Aspergillus* species and the clinical spectrum depends on the immune status of the infected host. Correlation of clinical spectrum of aspergillosis and immune status in various condition has been depicted in **Figure 1**.

Immunocompetent Patient*:* In immunocompetent individuals *Aspergillus* spp. remain colonized as a saprophytic fungus. *Aspergillus* spp. can colonize in pre-existing cavities due to bronchiectasis, tuberculosis, cavitary neoplasia or sarcoidosis and cause chronic non-invasive infections like chronic pulmonary aspergillosis (CPA) [18, 19].

Hyper responsive or Atopic Patient: A hypersensitive response in these individuals in various forms like Allergic bronchopulmonary aspergillosis (ABPA), severe asthma with fungal sensitization (SAFS) and allergic rhinitis [20]. This is commonly seen in patients with cystic fibrosis (CF) and poorly controlled or steroid-refractory asthma [20]. In cases of CF, inflammation of bronchial mucosa and abnormal mucus can result in fungal colonization and up to 10% patients develop sensitization to *A. fumigatus* [21]. This can further progress to ABPA suggesting the importance of testing such patients with markers of immune hyper-reactivity.

Immunocompromised Patient: IA is a dreaded, life-threatening disease with a high mortality ranging from 40–80% [22, 23]. It is commonly seen in are individuals with hematological malignancies such as acute leukemia; solid-organ and hematopoietic stem cell transplant patients; patients on prolonged corticosteroid or chemotherapy. Invasive pulmonary aspergillosis (IPA) is also reported in patients with history of influenza or coronavirus disease and those receiving broad-spectrum antibiotics [24, 25]. Genetic susceptibly to IA is also seen in patients

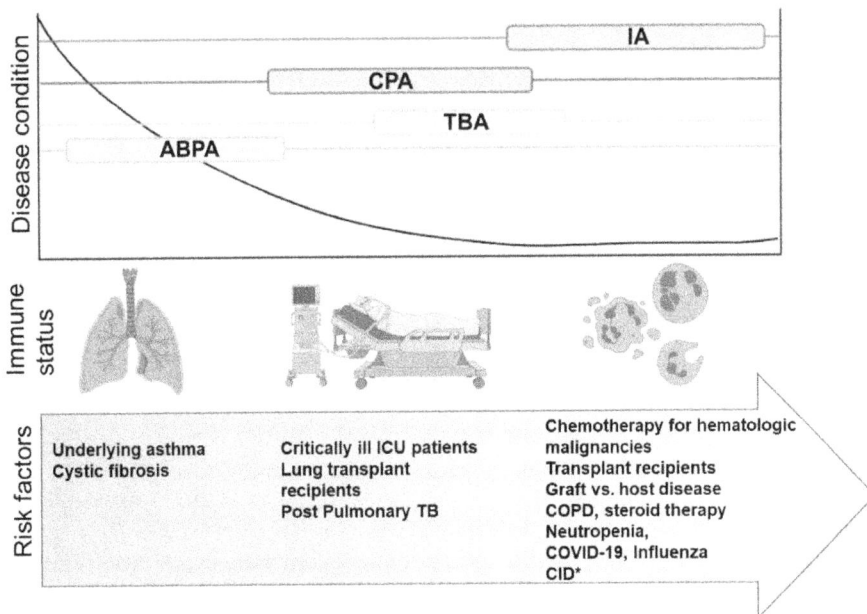

Figure 1.
Correction of clinical spectrum of Aspergillosis *and immune status in various condition. *CID: Congenital immunodeficiency disorders includes chronic granulomatous disease, CARD9 deficiency, leukocyte adhesion deficiency, Job's syndrome, pulmonary alveolar proteinosis.*

with congenital immune deficiencies like Caspase recruitment domain-containing protein-9 (CARD-9) deficiency and Chronic granulomatous disease [26, 27].

4. Pathogenesis

The range of ailments caused by *Aspergillus* depends on the host immune status. In atopic individuals the T helper 2 lymphocyte leads to hypersensitive response with increase in eosinophil counts and serum IgE levels. Formation of non-invasive aspergillomas is seen in CPA following repeated exposure to conidia in pre-existing cavitary lesions. IA is a destructive form of *Aspergillus*-related disease seen commonly in immunocompromised and critically ill patients.

5. Pathogen entry

The mode of reproduction in *Aspergillus* is predominantly asexual by formation of conidia (2–5 μm in size) which are ubiquitously present in the environment. These dormant conidia disperse in air easily due to their small size and common occurrence in soil, seeds and grains, decaying vegetation etc. and humans can inhale several hundred conidia per day. *Aspergillus* spp. are also found indoors in moisture damaged buildings both at homes and healthcare facilities [28] . There are therefore recommendations to avoid known sources of fungal proliferation (plants and flowers) in indoor places as they can serve as natural niches for fungal growth [29].

Conidia being small bypass the natural host nasal and bronchial defenses. The rodlet layer forms a hydrophobic layer outside conidia and protects it from host defenses and reach the lung alveoli. Natural defenses like mucociliary clearance and cough reflex are further compromised in intubated and mechanically ventilated patients. Also, the tracheal and bronchial epithelium is injured and provides easier passage for fungal conidia to the lower respiratory tract. Among healthy hosts, neutrophils and macrophages effectively clear the *Aspergillus* conidia. However, in immunocompromised patients, few conidia start swelling and become metabolically active after losing the outermost rodlet layer. These conidia, then germinate to produce fungal hyphae and cause a spectrum of invasive diseases.

6. Interaction with the innate immune system

The interaction of *Aspergillus* with cells of the innate immune system is depicted in **Figure 2**.

6.1 Soluble lung components

Various soluble factors found in the bronchopulmonary fluid are involved in *Aspergillus* defense including pathogen recognition receptors (PRRs) like C-type lectins, mannose binding ligand (MBL), Surfactant proteins (SP) – A and –D and pentraxin (PTX). These soluble factors enhance complement activation and phagocytosis of conidia, thus contributing to its clearance.

Although components of the complement system are predominant in serum they can also be found at lower levels in bronchial and alveolar fluid. Conidia and hyphae of *Aspergillus* species have been shown to bind to C3 followed by its cleavage to a ligand for phagocytic complement receptors iC3b. It has been reported

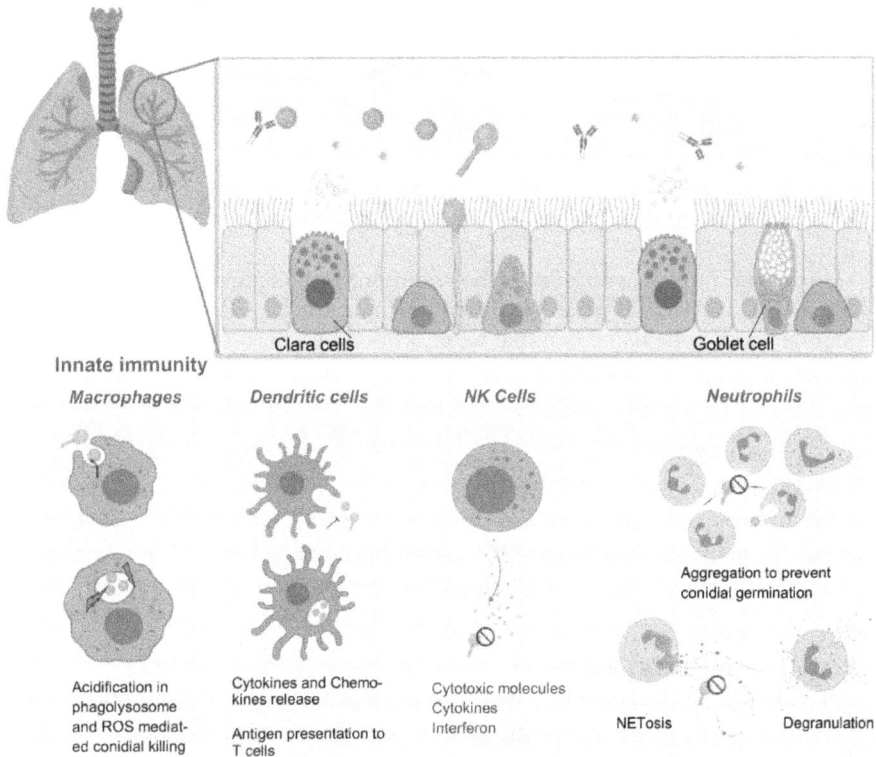

Figure 2.
Innate immune response to **Aspergillus** *infection. The conidia of* **Aspergillus** *spp. are inhaled and enter the lung where they encounter various soluble lung components including antibodies, complement factors and anti-microbial compounds. Those conidia which swell and undergoes germination further interact with a variety of innate immune cells including alveolar macrophages, dendritic cells, and NK cells. Conidial germination and development of hyphal forms is also prevented by neutrophils.*

that the common pathogens *A. fumigatus* and *A. flavus* bind to fewer C3 molecules compared to other to species making their complement-mediated phagocytosis and killing, less effective [30]. Hyphae and conidia from various *Aspergillus* spp. bind to alternative complement receptors like complement inhibitor factor H and the factor H family protein FFHL-1 which prevents complement cascade activation thereby protecting the fungus [31]. *A. fumigatus* and *A. flavus* have also been seen to produce a soluble complement-inhibitory factor which inhibits the activation of the alternative complement pathway [32]. This also acts as a defense mechanism of these species contributing to their overall pathogenesis.

6.2 Respiratory epithelial cells

The airway epithelial cells are the first cells to encounter inhaled *Aspergillus* conidia, which bind to it via sialic acid residues and subsequently modulate it. Other conidial proteins also mediate binding to fibrinogen, laminin and fibronectin which are all linked with lung injury indicating a role in adhesion and colonization [30]. A broad range of antimicrobial peptides of the defensin family are produced by the respiratory epithelial cells. Although the contribution of airway epithelial cells is less robust than that of the alveolar macrophages and germinating conidia and hyphae of *Aspergillus* are recognized by various PRRs on epithelial cells and subsequently assist in initiating pro-inflammatory response.

The proteases secreted by *A. fumigatus* cause desquamation and shrinkage of the respiratory epithelial cells along with actin cytoskeletal rearrangement with loss of cellular attachment and focal contact, thus assisting in invasion by germinating hyphae [33]. Secondary metabolites like gliotoxin, fumagillin, helvolic acid, verruculogen also damage airway epithelium and interfere with mucocilliary clearance [30, 34].

6.3 Pathogen recognition by innate immune cells

The recognition of *Aspergillus* by host immune cells is mostly via the PRRs – TLR1, TLR2, TLR4, TLR6 and the C-type lectin receptor i.e. dectin-1 [35]. TLR2 recognizes both hyphal and conidial form, while TLR4 recognizes only the hyphal morphology [36, 37]. The protective role of TLR4 mediated immune recognition has been seen in allogeneic hematopoietic stem cell transplant patients where it is observed that TLR4 polymorphisms are associated with IA [38]. The critical role of TLR6 in regulation of allergic inflammatory response in chronic fungal-induced asthma was studied by Moreira et al. in mice and the absence of TLR6 was found to be associated with less production of IL-23 and Th17 responses causing exacerbation of asthma [39]. Interestingly, the inflammatory response to *A. fumigatus* is intact in alveolar macrophages even in the setting of TLR2 deficiency and mice with defects in TLR2/TLR4 or its downstream effectors (like MyD88) have higher susceptibility to *A. fumigatus* lung infection, only in the setting of neutropenia [40–42].

Dectin – 1 is also an important PRR recognizing beta (1,3)-glucan on *Aspergillus* in both immunosuppressed and immunocompetent hosts. Although beta (1,3)-glucan is usually masked by the rodlet layer on resting conidia, the conidial swelling on entry in host epithelium exposes it, causing dectin – 1 mediated recognition and phagocytosis. Macrophages stimulation by *A. fumigatus* conidia increases intracellular PRR expression as well eg. Nucleotide-binding oligomerization domain (NOD) proteins ((NOD1 and NOD2) followed by production of proinflammatory cytokines which contribute to innate immune response [43].

6.4 Alveolar macrophages

Alveolar macrophages recognize and phagocytose fungal (1,3)-glucan bound to dectin-1. Internalization of conidia occurs within 2 hours and then conidial swelling begins [44]. This is an important requirement for induction of reactive oxygen species (ROS) production by the macrophage. Kinetic studies indicate that maximum ROS production occurs after 3 hours of phagocytosis resulting in fungistatic inhibition of germ tube formation due to which conidia are unable to germinate [44]. In immunosuppressed mice, although corticosteroid intake does not directly affect the internalization of conidia by alveolar macrophages there is impaired killing of *A. fumigatus* conidia due to defective production of ROS thereby increasing susceptibility to IA [44, 45]. The exact mechanisms of conidial killing by ROS are unknown and could be via direct toxicity or by acting as a cofactor for other phagolysosomal toxic molecules like elastase, cathepsins, proteases and chitinases [46]. In addition to phagolysosome acidification, phosphatidylinositol (PI) 3-kinase activity is also an important requirement for proper killing of conidia [47].

Neutrophils and macrophages produce nitric oxide (NO) and reactive nitrogen intermediates (RNI) that can also contribute to conidial killing. However, the expression of nitrogen oxidative species (NOS) which is seen in classically activated or M1 macrophages does not have much effect on conidial killing. A study by Lapp et al., reported that in *A. fumigatus* genes encoding flavohemoglobins (*FhpA* and

FhpB) which converts NO to nitrate and *S*-nitrosoglutathione reductase (*GnoA*) which reduce *S*-nitrosoglutathione to ammonium and glutathione disulphide are observed [48]. Although, these genes play a major role in detoxification of host derived RNI, they were not found to be essential for virulence.

Following macrophage phagocytosis, dihydroxynapthalene-melanin (DHN-melanin) of *A. fumigatus* prevents the phagolysosome acidification allowing conidial germination. However, *A. terreus* conidia lack the genes for DHN-melanin synthesis and instead produce a different type of melanin, i.e., Asp-melanin [49]. Although Asp-melanin does not impede acidification of phagolysosome it hampers phagocytosis and contributes to the survival and long-term persistence of *A. terreus* even in acidic environment.

In a study by Bhatia et al., alveolar macrophages were found to express Arginase 1 (Arg1) a key marker of alternatively activated macrophages (AAMs)/M2 macrophages after infection by *A. fumigatus* [50]. These macrophages efficiently phagocytose conidia and play a crucial role in pathogen clearance. The activation of macrophages is also followed by translocation of mitogen-activated protein kinases (MAPKs) to the nucleus where they phosphorylate the transcription factor NF-kappa B, thus activating a pro-inflammatory immune response.

6.5 Neutrophils

Neutrophils are professional phagocytes playing a pivotal role in innate immunity. Neutrophil recruitment is essential for effective *Aspergillus* clearing as they attack the germinating conidia and extracellular hyphae which have escaped macrophage surveillance. Neutrophils utilize TLR2, TLR4 and dectin-1, to identify and respond to *Aspergillus*. It can also be recognized directly by the complement receptor 3 (CD3, i.e., CD 11b/CD18), antigen–antibody complex detection by the Fcγ receptors (FcγR) or indirectly by opsonisation by various soluble components in lung environment.

In a study by Braem et al., higher deposition of the serum C3b was reported on germ tubes and swollen conidia compared to dormant conidia [51]. Also, patchy deposition of both C3b and immunoglobulin G (IgG) is seen over dormant conidia compared to uniform deposits on other morphotypes.

The release of chemotactic molecules, like C5a, increases migration of neutrophil to the infection site. The soluble mammalian extracellular β-galactose-binding lectin, galectin-3 is released in infected host tissues and facilitates neutrophil recruitment to the site of *A. fumigatus* infection by directly stimulating neutrophil motility in addition to exhibiting with both antimicrobial and immunomodulatory activities [52].

Neutrophil mediated killing involves both oxidative killing by NADPH oxidase which generates superoxide and myeloperoxidase and non-oxidative granule proteins containing various compounds with antimicrobial activity e.g., defensins, serine proteases, lysozyme, pentraxin-3 and lactoferrin [53]. Neutrophils attach to hyphae, spread over their surfaces, and degranulate thereby damaging the fungal hyphae. Neutrophils form aggregates in the lung and restrict conidial germination via lactoferrin mediated sequestration of iron [54]. Also, neutrophils produce lipocalin-1, which sequesters fungal siderophores thereby inhibiting fungal growth [55].

Another neutrophil dependent defense is the formation of neutrophil extracellular traps (NETs). Conidia and germ tubes of the *A. fumigatus* have been shown to trigger the formation of NETs. Pathogens in contact with the NETs become immobilized, limiting the spread of the infection. Calprotectin, a chelator of Zn^{2+} and Mn^{2+} ions is also produced by neutrophils and is associated with the *Aspergillus*-induced

NETs [56, 57]. Thus, in view of the important role that neutrophils play against *Aspergillus*, it is no surprize that patients with qualitative or quantitative defects in the neutrophils experience a greater risk of IA. It is worth mentioning however, that neutrophils may act as double-edged swords, since these are needed for fungal eradication but can also cause further lung injury by release of proteases and ROS. Thus, stringent regulatory mechanisms are essential to balance the protective activity and immunopathological responses for efficient control of the *Aspergillus.*

6.6 Natural killer cells

There is growing evidence suggesting the role of NK cells in immune response against *Aspergillus* spp. Direct antifungal activity via cytotoxic molecules like perforin and NK cell derived cytokines and interferon modulate the activation of other immune cells. *A. fumigatus* activates NK cells resulting in the production of low-levels of TNF-α, IFN-γ and lytic granules and release of fungal DNA [58]. These cells are a major source of early IFN-gamma production in the lungs of neutropenic patient with IA causing higher expression of IFN-inducible chemokines and subsequently enhancing macrophage antimicrobial effects. Studies in mice-models also suggest a critical role of NK cells in the pulmonary clearance of *A. fumigatus* [59].

Interestingly, in a study by Santiago et al., down-regulation of NK cell activating receptors NKG2D and NKp46 and a failure of full granule release was observed on contact of NK cells with *A. fumigatus* hyphae [59]. They also reported *A. fumigatus*-mediated NK cell immune-paresis which reduces cytokine-mediated response causing immune evasion during pulmonary aspergillosis [59]. Characterization of the clinical impact of NK cells in antifungal host immune response is still in its nascent stage as it involves complex interplay between multiple arms of the immune system [60].

6.7 Dendritic cells

Dendritic cells (DCs) bridge the innate and adaptive immune responses. They not only sense and patrol the lung environment but also initiate host response by antigen presentation which primes the T cell responses and causes cytokine secretion. Immature DCs are phagocytic and constantly perform surveillance of the lung environment while expressing PRRs like TLR 1, 2, 3, 4, 6 and Dectin-1 on cell surface that recognize various pathogen-associated molecular patterns (PAMPs). After phagocytosis, *A. fumigatus* conidia have been reported to escape from DCs, whereas some species like *A. terreus* persist with long-term survival, protecting them from anti-fungal action [49].

Typically, DCs are of two types, the plasmacytoid (pDCs) which are IFNα (type I interferon)-producing cells with a significant role in antifungal response and Classical (cDCs) which remain in the lymphoid tissue and cross-present antigens to T cells [61]. There is considerable plasticity in the functional activity of pulmonary DCs depending on the morphology of invading fungus [62].

1. Although DCs internalize both conidial and hyphal form of *A. fumigatus,* internalization of conidia occurs by coiling phagocytosis while entry of hyphae occurs by zipper-type phagocytosis. Also, phagocytosis of conidia is via involvement of a C-type lectin receptor while CR3 together with FcγR mediate the entry of opsonized hyphae.

2. Cytokine production is also variable depending on the fungal morphotype as TNF-α response is seen to any fungal form, but IL-12 is produced on exposure to conidia, while IL-4/IL-10 upon phagocytosis of hyphae.

3. The pulmonary DC transport *Aspergillus* fungal forms to the draining lymph nodes and spleen followed by functional maturation and eventual degradation for efficient antigen presentation.

4. The DCs also direct both local and peripheral T helper cell in response to fungus.

7. Interaction with the adaptive immune system

The adaptive immune response to *Aspergillus* infection is depicted in **Figure 3**.

7.1 Role of T-cells

Antigen-specific Th1 cells are crucial sources of IFN-γ mediated protective immunity to *A. fumigatus* [18, 58]. Peripheral blood of healthy adult donors has been found to have *A. fumigatus* specific effector/memory CD4 T cells with Th1 phenotype [63, 64]. A Th17 phenotype is noted in lung-derived *Aspergillus*-specific T cells [65]. IL-22 is produced by Th17 cells and has shown to play a crucial role in regulating *Aspergillus* induced asthma [66]. Like neutrophils, Th17 responses represent a "double-edged sword". During pulmonary fungal infections, the Th17 cell usually display a highly pro-inflammatory profile, which is detrimental to the infected host.

The Th2 cell-mediated immune responses along with Th1 and Th17 induces chronic pulmonary inflammation and lead to significant lung damage [67, 68]. This

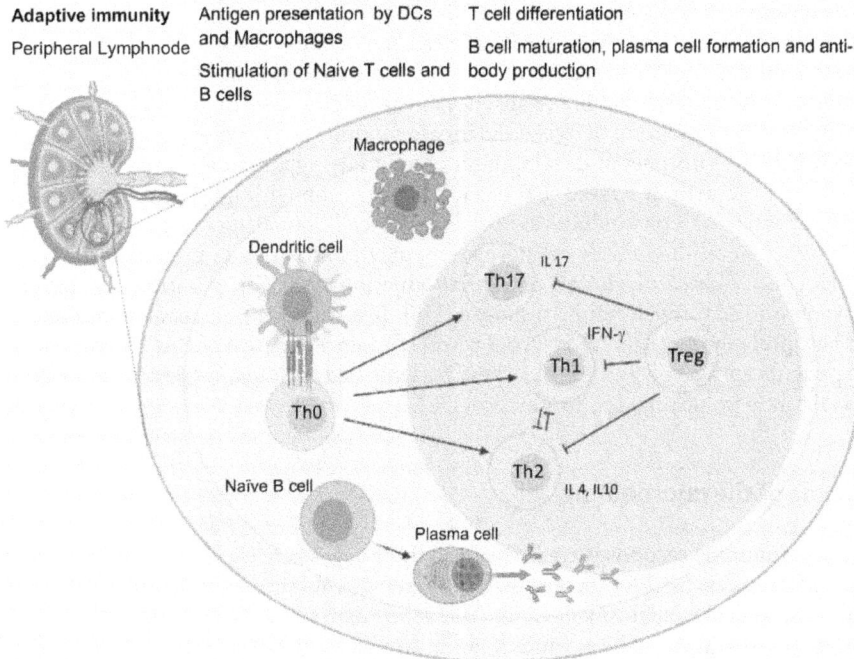

Figure 3.
Adaptive immune response to aspergillus *infection. Aspergillus spp. antigens are presented to naive T cells in peripheral lymphoid organs by dendritic cells and macrophages which further induces inflammation with coevolution of Th1, Th2, and Th17 response. B cells are also stimulated resulting in formation of anti-fungal antibody producing plasma cells.*

allows influx of macrophages followed by differentiation of both M1 and M2 sub-types [69]. These macrophages and T cells play a key role subsequently promoting extensive remodeling of medium- and small-sized pulmonary arteries. Pulmonary artery pathology including an increase in intimal area, smooth muscle proliferation, calcification of elastic membrane, and narrowed arterial lumens is seen in those with fatal asthma [70].

In healthy subjects, a strong Treg response has been seen as a part of the normal physiological T-cell repertoire which counterbalances the *A. fumigatus* specific T cells [71]. This intriguing finding raises the possibility that colonizing *A. fumigatus* may selectively promote Treg responses and subsequently limit antifungal immune activity. Activation of indoleamine 2,3- dioxygenase (IDO) as a regulator of infec-tion-linked tissue pathology is now being recognized as it acts via local tryptophan depletion, or generation of immunomodulatory metabolites. Interaction of TLRs with PAMPs induces IDO which regulates the inflammatory/anti-inflammatory status of the innate immune cell and modifies the local tissue microenvironment. There is also activation of GCN2, a T-cell stress-response kinase which senses amino acid starvation and impairs lymphocyte proliferation while enhancing polarization toward a Treg phenotype [72]. In patients of CF with ABPA, dysregulation of the IDO pathway is seen at both the genetic and transcriptional levels, leading to an imbalanced Th17/Treg with high Th2 polarization resulting in chronic inflamma-tion and significant lung damage in response to *A. fumigatus* [73].

7.2 Role of B-cells

In a study by Montagnoli et al., the role of B cells and antibodies in the genera-tion of antifungal immune resistance was studied in B cell-deficient (μMT) mice which were infected with *A. fumigatus* [74]. They reported that, although passive transfer of antibodies helped in fungal clearance, a compensatory increase in both innate and Th1-mediated resistance to infection was seen in μMT mice with asper-gillosis. This suggests that in the absence of opsonizing antifungal antibodies, the nature of the interaction between the innate immune cells and with fungi may be modified with subsequent development of long-lasting antifungal immunity [74].

Chen et al., demonstrated that basophil interaction with IgD bound antigens and activation of TLRs induces expression of B-cell-activating factor (BAFF), an important regulator of B-cell activation, proliferation, and immunoglobulin production. This results in IgG and IgE production by B cells, pointing to a role of basophils in adaptive immune responses [75]. In a study by Boita et al. stimulation of basophil membrane by *Aspergillus* resulted in upregulation of BAFF expression in patients with SAFS and ABPA. These patients had high IgE suggesting the role of basophils in polyclonal IgE production [76].

8. Role of the microbiome

Host immune responses are influenced by changes in the gut microbiome. Short-chain fatty acids (SCFAs) produced by the gut microbiome are recognized by innate immune cells like macrophages and neutrophils expressing G-coupled protein receptor GPR43 [77]. The gut microbiome also plays a crucial role in anti-*Aspergillus* host defense by coordinating lymphocyte subsets at the mucosal level in distant organs such as the lungs. Although, fungal microbiome compromise <0.1% of total microbiome, fungal cell components such as β-glucans may influence immune responses as perceived by their role in autoimmune diseases [78]. *In-vivo* studies in mice have revealed that intake of SCFA (propionate/butyrate) or supplementation of diet with fermentable fibers which increases SCFA producing bacteria, increases the generation of DCs and macrophages in the lung and bone marrow with increased phagocytic capacity [79–81]. These alterations also reduce the ability to prime cells toward Th2 responses lowering DC ability to induce *Aspergillus*-allergic inflammation [82].

The intestinal segmented filamentous bacterium (SFB) have been shown to induce Th17 cells producing IL-17 and IL-22 in the lamina propria of the gut and can even regulate pulmonary adaptive immune response by increasing Th17 responses in the lung [83, 84]. However, it is important to determine whether lung microbiome also has similar Th17-polarizing ability which can influence anti-*Aspergillus* host response.

It has also been observed that in germ-free mice, the absence of commensal gut microbiota leads to increase susceptibility to pulmonary viral infections. Hence, the gut microbiome can influence pulmonary immune responses by release of type 1 IFN [85, 86]. Intestinal colonization of microorganism is necessary for cytotoxic activity by NK-cell, CD8$^+$ T-cell clonal expansion, and production of specific antibodies [85].

Recently, innate lymphoid cells (ILCs) have emerged as an important cell population that has the capacity to synchronize microbiome-related immune regulation [87]. ILCs can express functional TLR2 which on stimulation induces IL-2 production, subsequently increasing the expression of IL-22, enhancing the allergic airway responses induced by *Aspergillus* spp [88]. It has also been observed that commensal bacterial limit the production of serum IgE levels which directly influences bone marrow - basophil precursors, leading to increased allergic airway responses [89].

The treatment of diseases like COPD with steroids and bronchodilators, may also alter the microbiome [90] which can subsequently increase the risk of colonization and infection by *Aspergillus* spp. In patients with Influenza, significant changes in the lung microbiome have been observed with a relative abundance of *Firmicutes* and *Proteobacteria* more specifically, *Pseudomonas* spp., which contributes to secondary invasive infections by *Aspergillus* spp. [91, 92]. Other factors like antibiotic exposure can also influence the micro-environment of the microbiome, which can affect the pulmonary immune responses to *Aspergillus* causing allergic airway diseases [93]. In patients with CF, interaction between fungal and bacterial pathogens and their biofilms may influence pathogenicity which can be observed by significant decrease in *Aspergillus* in the sputum on treatment with anti-pseudomonal antibiotics [94, 95].

9. Genetic susceptibility to aspergillosis

The genetic polymorphisms within pattern recognition receptors PRRs (*TLR1, TLR2, TLR4, TLR5, TLR6, TLR9, Dectin-1, Dectin-2, DC-SIGN, MASP, MBL, PTX-3* surfactant protein-A2 and plaminogen) cytokines (*IL1, IL10,* IFN- γ, *CXCL10, ARNT2,*) and their receptors (*CX3CR1* and IL-4Rα) is depicted in –**Table 2**.

10. Conclusion

The clinical spectrum of *Aspergillus* related infections depends on the host immune status ranging from allergic manifestations in immunocompetent atopic individuals to invasive disease in immunosuppressed individuals. Various components of the innate and adaptive immune system form an intricate network modulating host response to *Aspergillus* exposure. Many future studies are required to study

Gene	Function	SNP position	Disease condition	Reference
Pattern Recognition receptors (PRRs)				
TLR1	TLR1 forms heterodimer with TLR2 and facilitate the fungicidal activity by various oxidative pathways	239 C/G [80 R/T] 743 A/G [248 S/N] 1063 A/G	IA	[96]
TLR 2	TLR-2 act as PRR for *Aspergillus* spp. Antigens and activate innate immune cells. Further downstream signaling via TLR2 promote the fungicidal activity by various oxidative pathways which lead to proinflammatory cytokines release.	Arg753Gln (G + 2258A) polymorphism affects the TIR domain of TLR-2 and impairs its functional activity.	IA	[97]
TLR4	TLR4 promotes fungicidal activity	[299 D/G] 1363 C/T [399 I/T] 1063 A/G [299 D/G]	IA after HSCT [EORTC] CCPA	[38, 98, 99]
TLR5	TLR-5 induction causes increase in expression of pro-inflammatory cytokines	1174C T (STOP codon)	IA	[100]
TLR6	It promotes IL-23 release and a subsequent Th17 response.	745 C/T [249 S/P]	IA after HSCT [EORTC]	[96]
TLR9	It recognizes unmethylated CpG DNA and induces innate immune responses.	1237 C/T [Promotor]	ABPA	[98] [101]
Dectin-1	Dectin-1 is act as a PRR, which is present on myeloid cells surface and expressed by DCs and macrophages. It is specialized for recognition of β-1,3-glucan of fungal species. It leads to production of chemokines and cytokines and causes recruitment of neutrophil recruitment and ROS production.	Y238X polymorphism [Stop Codon Polymorphism]	IA	[102] [103] [104]
Dectin-2	Dectin-1 is act as a PRR, which is present on plasmacytoid dendritic cells (pDCs). It is specialized for recognition of α-mannans of fungal species. It leads to cytokine production, extracellular trap (pET) formation and ROS production.	(CLEC6A – A/G) [Intron] (CLEC6A - C/T) [Intron]	IPA	[104]
DC-SIGN	DC-SIGN is a CLR. It recognizes galactomannans.	336 A/G [promoter] c.898 A/G [3′-UTR] c.74928 C/T [3′-UTR] IVS2 + 11 G/C [Intron]	IPA	[104]

Gene	Function	SNP position	Disease condition	Reference
pentraxin (*PTX3)*	PTX3 is a soluble opsonin. It is produced by phagocytes that facilitates microbial recognition and phagocytosis of conidia.	+281A/G[Intron 1] +734A/C (D48A) [Exon 2] +1449A/G [Intron 2]	IA	[105] [106]
Mannose-binding lectin-associated serine protease (MASP2)	MASP binds directly to *Aspergillus fumigatus* and promote complement activation and phagocytosis	380 A/C [D120G]	IA	[107]
MBL	MBL is a soluble PRR. It opsonizes the carbohydrate moieties of fungus and activates the lectin complement pathway using the MASPs and induces the release of proinflammatory cytokines.	868 C/T [52 C/R] 1011 A/G [Intron] 868 C/T [52 C/R]	CCPA ABPA CNPA	[108–113]
Plg	Plasminogen is produced by phagocytes that facilitates microbial recognition.	28904 A/G[a] [472 N/D]	IA	[114–116]
SFTPA2 surfactant protein-A2		1660 A/G [94 R/R] 1649 C/G [91 A/P] 1492 C/T [Intron]	ABPA	[117, 118]
Cytokines				
CXCL10	It is an 'inflammatory' chemokine. It binds to CXCR3 and mediate leukocytes recruitment such as eosinophils, T cells, NK cells and monocytes.	11101 C/Ta [Downstream] 1642 C/Ga [3′ UTR] 1101 A/Ga [Promotor]	IA	[119] [120] [121]
ARNT2	It regulates the activity and differentiation of phagocytic cells like macrophages and lymphocytes.	80732053 [Intron]	IA	[122]
IFN-γ	It promotes differentiation of Th1 response	1616 C/T[a] [Promotor] 1082 A/G [Promotor]	IA	[123]
IL-10	IL-10 plays a significant role in the development of atopy. It inhibits the activity of Th1 cells, NK cells, and macrophages which are essential for clearance of fungus.	2068 C/G[a] [Intron] 1082 A/G [Promotor] 1082 A/G – 819 C/T – 592 A/C [Promotor] 1082 A/G [Promotor]	IA ABPA	[124] [125] [126]
IL-4R alpha	IL-4 released by T cells binds to the IL-4 receptor (IL-4R) on B cells resulting in B cell proliferation and IgE isotype switching.	4679 A/C/G/T [75 I/L/F/V]	ABPA	[127]
Cytokine's receptors				
TNFR2 TNF receptor type 2	TNFR2 (p75) receptor is expressed by T regulatory cells for survival during clonal expansion.	322 [Promotor]	IPA	[107]

Gene	Function	SNP position	Disease condition	Reference
Interferon regulatory factor - 4 (*IRF4*)	It regulates the NFκB pathway and cell proliferation and modulates the differentiation of different DC and Th17-mediated immune responses against *Aspergillus fumigatus*.	rs12203592	IA	[128]
CX3CR1	Modulates the interaction of fungal pathogens with immune phagocytes.	39286825 [Intron] 39293757 [Intron]	IA	[122]

TLR-Toll-like receptor, IL – Interleukin, PRR – Pathogen Recognition Receptor, *Th – T helper cells, DC-SIGN - Dendritic Cell-Specific Intercellular adhesion molecule-3-Grabbing Non-integrin, PTX3- Pentraxin, MASP2 - Mannose-binding lectin-associated serine protease, MBL - Mannose-binding lectin, CXCL - chemokine (C-X-C motif) ligand, ARNT2 - Aryl hydrocarbon receptor nuclear translocator 2, IL-4R alpha - Interleukin 4 receptor alpha, TNFR2 - TNF receptor type 2, IRF4 Interferon regulatory factor - 4, CX3CR1 - CX3C chemokine receptor 1, IA- invasive aspergillosis, IPA- invasive pulmonary aspergillosis, CCPA- Chronic cavitary pulmonary aspergillosis, ABPA – Allergic bronchopulmonary aspergillosis, CNPA – Chronic necrotizing pulmonary aspergillosis, HSCT-Hematopoietic stem cell transplantation, EORTC- European Organization for Research and Treatment of Cancer.*

Table 2.
Summary of immune system related genes mediating susceptibility to aspergillosis.

the association and impact of the complex interactions between the gut/pulmonary microbiome and the immune system in *Aspergillus*-related diseases. An understanding of the immune pathogenesis of aspergillosis can help in the development of strategies targeting *Aspergillus* itself as well as pulmonary or systemic immunity by influencing the host immune system, the microbiome and/or its metabolites.

Acknowledgements

All artworks are original and was prepared using the trial version of the online Biorender software.

Author details

Shreya Singh, Rimjhim Kanaujia and Shivaprakash M. Rudramurthy[*]
Department of Medical Microbiology, Postgraduate Institute of Medical Education and Research, Chandigarh, India

*Address all correspondence to: mrshivprakash@yahoo.com

IntechOpen

References

[1] Chakrabarti A: *Fungal infections in Asia : Eastern frontier of mycology.* Elsevier India 2014.

[2] Chakrabarti A, Chatterjee SS, Das A, Shivaprakash MR: Invasive aspergillosis in developing countries. *Medical Mycology* 2011:S35–S47.

[3] Holdom MD, Hay RJ, Hamilton AJ: Purification, n-terminal amino acid sequence and partial characterization of a Cu,Zn superoxide dismutase from the pathogenic fungus *Aspergillus fumigatus*. *Free Radic Res* 1995, 22:519-531.

[4] Kothary MH, Chase T, Macmillan JD: Correlation of elastase production by some strains of *Aspergillus fumigatus* with ability to cause pulmonary invasive aspergillosis in mice. *Infect Immun* 1984, 43:320-325.

[5] Ramesh M V, Sirakova T, Kolattukudy PE: Isolation, characterization, and cloning of cDNA and the gene for an elastinolytic serine proteinase from *Aspergillus flavus*. *Infect Immun* 1994, 62:79-85.

[6] Sirakova TD, Markaryan A, Kolattukudy PE: Molecular cloning and sequencing of the cDNA and gene for a novel elastinolytic metalloproteinase from *Aspergillus fumigatus* and its expression in *Escherichia coli*. *Infect Immun* 1994, 62:4208-4218.

[7] Reichard U, Eiffert H, Rüchel R: Purification and characterization of an extracellular aspartic proteinase from *Aspergillus fumigatus*. *Med Mycol* 1994, 32:427-436.

[8] Calera JA, Paris S, Monod M, Hamilton AJ, Debeaupuis JP, Diaquin M, López-Medrano R, Leal F, Latgé JP: Cloning and disruption of the antigenic catalase gene of *Aspergillus fumigatus*. *Infect Immun* 1997, 65:4718-4724.

[9] Krappmann S, Bignell EM, Reichard U, Rogers T, Haynes K, Braus GH: The *Aspergillus fumigatus* transcriptional activator CpcA contributes significantly to the virulence of this fungal pathogen. *Mol Microbiol* 2004, 52:785-799.

[10] Mullbacher A, Eichner RD: Immunosuppression in vitro by a metabolite of a human pathogenic fungus. *Proc Natl Acad Sci* 1984, 81:3835-3837.

[11] Mullbacher A, Waring P, Eichner Rd: Identification of an agent in cultures of *Aspergillus fumigatus* displaying anti-phagocytic and immunomodulating activity in vitro. *Microbiology* 1985, 131:1251-1258.

[12] Latgé JP, Moutaouakil M, Debeaupuis JP, Bouchara JP, Haynes K, Prévost MC: The 18-kilodalton antigen secreted by *Aspergillus fumigatus*. *Infect Immun* 1991, 59:2586-2594.

[13] Paris S, Monod M, Diaquin M, Lamy B, Arruda LK, Punt PJ, Latgé JP: A transformant of *Aspergillus fumigatus* deficient in the antigenic cytotoxin ASPFI. *FEMS Microbiol Lett* 1993, 111:31-36.

[14] Robens JF, Richard JL: Aflatoxins in Animal and Human Health. *Rev Environ Contam Toxicol.* 1992:69-94.

[15] Eissenberg LG, Schlesinger PH, Goldman WE: Phagosome-Lysosome Fusion in P388D1 macrophages infected with *Histoplasma capsulatum*. *J Leukoc Biol* 1988, 43:483-491.

[16] Hermanowski-Vosatka A, Detmers PA, Götze O, Silverstein SC, Wright SD: Clustering of ligand on the surface of a particle enhances adhesion to receptor-bearing cells. *J Biol Chem* 1988, 263:17822-7.

[17] Valsecchi I, Dupres V, Stephen-Victor E, Guijarro JI, Gibbons J, Beau R, Bayry J, Coppee JY, Lafont F, et al.: Role of Hydrophobins in *Aspergillus fumigatus*. *J Fungi* (Basel). 2017, 24;4:2.

[18] Alastruey-Izquierdo A, Cadranel J, Flick H, Godet C, Hennequin C, Hoenigl M, Kosmidis C, Lange C, Munteanu O, Page I, et al.: Treatment of Chronic Pulmonary Aspergillosis: Current Standards and Future Perspectives. *Respiration* 2018, 96:159-170.

[19] Denning DW, Cadranel J, Beigelman-Aubry C, Ader F, Chakrabarti A, Blot S, Ullmann AJ, Dimopoulos G, Lange C: Chronic pulmonary aspergillosis: Rationale and clinical guidelines for diagnosis and management. *Eur Respir J* 2016, 47:45-68.

[20] Agarwal R, Chakrabarti A, Shah A, Gupta D, Meis JF, Guleria R, Moss R, Denning DW: Allergic bronchopulmonary aspergillosis: Review of literature and proposal of new diagnostic and classification criteria. *Clin Exp Allergy* 2013, doi:10.1111/cea.12141.

[21] Singh M, Paul N, Singh S, Nayak GR: Asthma and Fungus: Role in Allergic Bronchopulmonary Aspergillosis (ABPA) and Other Conditions. *Indian J Pediatr* 2018, 85:899-904.

[22] Azie N, Neofytos D, Pfaller M, Meier-Kriesche HU, Quan SP, Horn D: The PATH (Prospective Antifungal Therapy) Alliance® registry and invasive fungal infections: Update 2012. *Diagn Microbiol Infect Dis* 2012, 73:293-300.

[23] Brown GD, Denning DW, Gow NAR, Levitz SM, Netea MG, White TC: Hidden Killers: Human Fungal Infections. *Sci Transl Med* 2012, 4:165rv13-165rv13.

[24] Alangaden GJ, Wahiduzzaman M, Chandrasekar PH: Aspergillosis: The Most Common Community-Acquired Pneumonia with Gram-Negative Bacilli as Copathogens in Stem Cell Transplant Recipients with Graft-versus-Host Disease. *Clin Infect Dis* 2002, 35:659-664.

[25] Schauwvlieghe AFAD, Rijnders BJA, Philips N, Verwijs R, Vanderbeke L, Van Tienen C, Lagrou K, Verweij PE, Van de Veerdonk FL, Gommers D, et al.: Invasive aspergillosis in patients admitted to the intensive care unit with severe influenza: a retrospective cohort study. *Lancet Respir Med* 2018, 6:782-792.

[26] Drummond RA, Franco LM, Lionakis MS: Human CARD9: A Critical Molecule of Fungal Immune Surveillance. *Front Immunol* 2018, 9.

[27] Hodiamont CJ, Dolman KM, Ten berge IJM, Melchers WJG, Verweij PE, Pajkrt D: Multiple-azole-resistant *Aspergillus fumigatus* osteomyelitis in a patient with chronic granulomatous disease successfully treated with long-term oral posaconazole and surgery. *Med Mycol* 2009, 47:217-220.

[28] Mousavi B, Hedayati MT, Hedayati N, Ilkit M, Syedmousavi S: *Aspergillus* species in indoor environments and their possible occupational and public health hazards. *Curr Med Mycol* 2016, 2:36-42.

[29] Hedayati MT, Mohseni-Bandpi A, Moradi S: A survey on the pathogenic fungi in soil samples of potted plants from Sari hospitals, Iran. *J Hosp Infect* 2004, 58:59-62.

[30] Dagenais TRT, Keller NP: Pathogenesis of *Aspergillus fumigatus* in invasive aspergillosis. *Clin Microbiol Rev* 2009, 22:447-465.

[31] Vogl G, Lesiak I, Jensen DB, Perkhofer S, Eck R, Speth C,

Lass-Flörl C, Zipfel PF, Blom AM, Dierich MP, et al.: Immune evasion by acquisition of complement inhibitors: The mould *Aspergillus* binds both factor H and C4b binding protein. *Mol Immunol* 2008, 45:1485-1493.

[32] Washburn RG, DeHart DJ, Agwu DE, Bryant-Varela BJ, Julian NC: *Aspergillus fumigatus* complement inhibitor: production, characterization, and purification by hydrophobic interaction and thin-layer chromato graphy. *Infect Immun* 1990, 58.

[33] Kogan TV, Jadoun J, Mittelman L, Hirschberg K, Osherov N: Involvement of Secreted *Aspergillus fumigatus* Proteases in Disruption of the Actin Fiber Cytoskeleton and Loss of Focal Adhesion Sites in Infected A549 Lung Pneumocytes. *J Infect Dis* 2004, 189:1965-1973.

[34] Arias M, Santiago L, Vidal-García M, Redrado S, Lanuza P, Comas L, Domingo MP, Rezusta A, Gálvez EM: Preparations for invasion: Modulation of host lung immunity during pulmonary aspergillosis by gliotoxin and other fungal secondary metabolites. *Front Immunol* 2018, 9:2549.

[35] Sales-Campos H, Tonani L, Cardoso CRB, Kress MRVZ: The immune interplay between the host and the pathogen in *Aspergillus fumigatus* lung infection. *Biomed Res Int* 2013, 2013.

[36] Rosentul DC, Delsing CE, Jaeger M, Plantinga TS, Oosting M, Costantini I, Venselaar H, Joosten LAB, van der Meer JWM, Dupont B, et al.: Gene polymorphisms in pattern recognition receptors and susceptibility to idiopathic recurrent vulvovaginal candidiasis. *Front Microbiol* 2014, 5:483.

[37] Netea MG, Warris A, Van Der Meer JWM, Fenton MJ, Verver-Janssen TJG, Jacobs LEH, Andresen T, Verweij PE, Kullberg BJ:

Aspergillus fumigatus evades immune recognition during germination through loss of toll-like receptor-4-mediated signal transduction. *J Infect Dis* 2003, 188:320-326.

[38] Bochud P-Y, Chien JW, Marr KA, Leisenring WM, Upton A, Janer M, Rodrigues SD, Li S, Hansen JA, Zhao LP, et al.: Toll-like Receptor 4 Polymorphisms and Aspergillosis in Stem-Cell Transplantation. *N Engl J Med* 2008, 359:1766-1777.

[39] Moreira AP, Cavassani KA, Ismailoglu UB, Hullinger R, Dunleavy MP, Knight DA, Kunkel SL, Uematsu S, Akira S, Hogaboam CM: The protective role of TLR6 in a mouse model of asthma is mediated by IL-23 and IL-17A. *J Clin Invest* 2011, 121:4420-4432.

[40] Bretz C, Gersuk G, Knoblaugh S, Chaudhary N, Randolph-Habecker J, Hackman RC, Staab J, Marr KA: MyD88 signaling contributes to early pulmonary responses to *Aspergillus fumigatus*. *Infect Immun* 2008, 76:952-958.

[41] Bellocchio S, Montagnoli C, Bozza S, Gaziano R, Rossi G, Mambula SS, Vecchi A, Mantovani A, Levitz SM, Romani L: The Contribution of the Toll-Like/IL-1 Receptor Superfamily to Innate and Adaptive Immunity to Fungal Pathogens In Vivo. *J Immunol* 2004, 172:3059-3069.

[42] Dubourdeau M, Athman R, Balloy V, Huerre M, Chignard M, Philpott DJ, Latgé J-P, Ibrahim-Granet O: *Aspergillus fumigatus* Induces Innate Immune Responses in Alveolar Macrophages through the MAPK Pathway Independently of TLR2 and TLR4. *J Immunol* 2006, 177:3994-4001.

[43] Li ZZ, Tao LL, Zhang J, Zhang HJ, Qu JM: Role of NOD2 in regulating the immune response to *Aspergillus*

fumigatus. Inflamm Res 2012, 61: 643-648.

[44] Philippe B, Ibrahim-Granet O, Prévost MC, Gougerot-Pocidalo MA, Perez MS, Van der Meeren A, Latgé JP: Killing of *Aspergillus fumigatus* by alveolar macrophages is mediated by reactive oxidant intermediates. *Infect Immun* 2003, 71:3034-3042.

[45] De Castro CMMB, Manhães De Castro R, Fernandes De Medeiros A, Queirós Santos A, Ferreira E Silva WT, De Lima Filho JL: Effect of stress on the production of O2- in alveolar macrophages. *J Neuroimmunol* 2000, 108:68-72.

[46] Reeves EP, Lu H, Jacobs HL, Messina CGM, Bolsover S, Gabellall G, Potma EO, Warley A, Roes J, Segal AW: Killing activity of neutrophils is mediated through activation of proteases by K+ flux. *Nature* 2002, 416:291-297.

[47] Ibrahim-Granet O, Philippe B, Boleti H, Boisvieux-Ulrich E, Grenet D, Stern M, Latgé JP: Phagocytosis and intracellular fate of *Aspergillus fumigatus* conidia in alveolar macrophages. *Infect Immun* 2003, 71:891-903.

[48] Lapp K, Vödisch M, Kroll K, Strassburger M, Kniemeyer O, Heinekamp T, Brakhage AA: Characterization of the *Aspergillus fumigatus* detoxification systems for reactive nitrogen intermediates and their impact on virulence. *Front Microbiol* 2014, 5:469.

[49] Hsieh SH, Kurzai O, Brock M: Persistence within dendritic cells marks an antifungal evasion and dissemination strategy of *Aspergillus terreus. Sci Rep* 2017, 7:1-11.

[50] Bhatia S, Fei M, Yarlagadda M, Qi Z, Akira S, Saijo S, Iwakura Y, van Rooijen N, Gibson GA, St. Croix CM, et al.: Rapid host defense against

Aspergillus fumigatus involves alveolar macrophages with a predominance of alternatively activated phenotype. *PLoS One* 2011, 6.

[51] Braem SGE, Rooijakkers SHM, van Kessel KPM, de Cock H, Wösten HAB, van Strijp JAG, Haas P-JA: Effective Neutrophil Phagocytosis of *Aspergillus fumigatus* Is Mediated by Classical Pathway Complement Activation. *J Innate Immun* 2015, 7:364-374.

[52] Snarr BD, St-Pierre G, Ralph B, Lehoux M, Sato Y, Rancourt A, Takazono T, Baistrocchi SR, Corsini R, Cheng MP, et al.: Galectin-3 enhances neutrophil motility and extravasation into the airways during *Aspergillus fumigatus* infection. *PLOS Pathog* 2020, 16:e1008741.

[53] Feldmesser M: Role of neutrophils in invasive aspergillosis. *Infect Immun* 2006, 74:6514-6516.

[54] Gazendam RP, van Hamme JL, Tool ATJ, Hoogenboezem M, van den Berg JM, Prins JM, Vitkov L, van de Veerdonk FL, van den Berg TK, Roos D, et al.: Human Neutrophils Use Different Mechanisms To Kill *Aspergillus fumigatus* Conidia and Hyphae: Evidence from Phagocyte Defects. *J Immunol* 2016, 196:1272-1283.

[55] Leal SM, Roy S, Vareechon C, Carrion S de J, Clark H, Lopez-Berges MS, diPietro A, Schrettl M, Beckmann N, Redl B, et al.: Targeting Iron Acquisition Blocks Infection with the Fungal Pathogens *Aspergillus fumigatus* and *Fusarium oxysporum. PLoS Pathog* 2013, 9.

[56] Clark HL, Jhingran A, Sun Y, Vareechon C, de Jesus Carrion S, Skaar EP, Chazin WJ, Calera JA, Hohl TM, Pearlman E: Zinc and Manganese Chelation by Neutrophil S100A8/A9 (Calprotectin) Limits Extracellular *Aspergillus fumigatus*

Hyphal Growth and Corneal Infection .
J Immunol 2016, 196:336-344.

[57] McCormick A, Heesemann L, Wagener J, Marcos V, Hartl D, Loeffler J, Heesemann J, Ebel F: NETs formed by human neutrophils inhibit growth of the pathogenic mold *Aspergillus fumigatus. Microbes Infect* 2010, 12:928-936.

[58] Park SJ, Hughes MA, Burdick M, Strieter RM, Mehrad B: Early NK Cell-Derived IFN-γ Is Essential to Host Defense in Neutropenic Invasive Aspergillosis. *J Immunol* 2009, 182:4306-4312.

[59] Santiago V, Rezvani K, Sekine T, Stebbing J, Kelleher P, Armstrong-James D: Human NK Cells Develop an Exhaustion Phenotype During Polar Degranulation at the *Aspergillus fumigatus* Hyphal Synapse. *Front Immunol* 2018, 9:2344.

[60] Zhang X, He D, Gao S, Wei Y, Wang L: *Aspergillus fumigatus* enhances human NK cell activity by regulating M1 macrophage polarization. *Mol Med Rep* 2019, 20:1241-1249.

[61] Reizis B, Bunin A, Ghosh HS, Lewis KL, Sisirak V: Plasmacytoid dendritic cells: Recent progress and open questions. *Annu Rev Immunol* 2011, 29:163-183.

[62] Bozza S, Gaziano R, Spreca A, Bacci A, Montagnoli C, di Francesco P, Romani L: Dendritic Cells Transport Conidia and Hyphae of *Aspergillus fumigatus* from the Airways to the Draining Lymph Nodes and Initiate Disparate Th Responses to the Fungus. *J Immunol* 2002, 168:1362-1371.

[63] Beck O, Topp MS, Koehl U, Roilides E, Simitsopoulou M, Hanisch M, Sarfati J, Latgé JP, Klingebiel T, Einsele H, et al.: Generation of highly purified and functionally active human TH1 cells

against *Aspergillus fumigatus. Blood* 2006, 107:2562-2569.

[64] Vogel K, Pierau M, Arra A, Lampe K, Schlueter D, Arens C, Brunner-Weinzierl MC: Developmental induction of human T-cell responses against *Candida albicans* and *Aspergillus fumigatus. Sci Rep* 2018, 8:16904.

[65] Jolink H, de Boer R, Hombrink P, Jonkers RE, van Dissel JT, Falkenburg JHF, Heemskerk MHM: Pulmonary immune responses against *Aspergillus fumigatus* are characterized by high frequencies of IL-17 producing T-cells. *J Infect* 2017, 74:81-88.

[66] Lilly LM, Gessner MA, Dunaway CW, Metz AE, Schwiebert L, Weaver CT, Brown GD, Steele C: The β-Glucan Receptor Dectin-1 Promotes Lung Immunopathology during Fungal Allergy via IL-22. *J Immunol* 2012, 189:3653-3660.

[67] Murdock BJ, Shreiner AB, McDonald RA, Osterholzer JJ, White ES, Toews GB, Huffnagle GB: Coevolution of TH1, TH2, and TH17 responses during repeated pulmonary exposure to *Aspergillus fumigatus* conidia. *Infect Immun* 2011, 79:125-135.

[68] Shreiner AB, Murdock BJ, Akha AAS, Falkowski NR, Christensen PJ, White ES, Hogaboam CM, Huffnagle GB: Repeated exposure to *Aspergillus fumigatus* conidia results in CD4 + T cell-dependent and -independent pulmonary arterial remodeling in a mixed th1/th2/th17 microenvironment that requires interleukin-4 (IL-4) and IL-10. *Infect Immun* 2012, 80:388-397.

[69] Arora S, Olszewski MA, Tsang TM, McDonald RA, Toews GB, Huffnagle GB: Effect of cytokine interplay on macrophage polarization during chronic pulmonary infection with *Cryptococcus neoformans. Infect Immun* 2011, 79:1915-1926.

[70] Shiang C, Mauad T, Senhorini A, De Araújo BB, Ferreira DS, Da Silva LFF, Dolhnikoff M, Tsokos M, Rabe KF, Pabst R: Pulmonary periarterial inflammation in fatal asthma. *Clin Exp Allergy* 2009, 39:1499-1507.

[71] Dewi IMW, van de Veerdonk FL, Gresnigt MS: The multifaceted role of T-helper responses in host defense against *Aspergillus fumigatus. J Fungi* 2017, 3.

[72] Munn DH, Sharma MD, Baban B, Harding HP, Zhang Y, Ron D, Mellor AL: GCN2 kinase in T cells mediates proliferative arrest and anergy induction in response to indoleamine 2,3-dioxygenase. *Immunity* 2005, 22:633-642.

[73] Iannitti RG, Carvalho A, Cunha C, De Luca A, Giovannini G, Casagrande A, Zelante T, Vacca C, Fallarino F, Puccetti P, et al.: Th17/Treg imbalance in murine cystic fibrosis is linked to indoleamine 2,3-dioxygenase deficiency but corrected by kynurenines. *Am J Respir Crit Care Med* 2013, 187:609-620.

[74] Montagnoli C, Bozza S, Bacci A, Gaziano R, Mosci P, Morschhäuser J, Pitzurra L, Kopf M, Cutler J, Romani L: A role for antibodies in the generation of memory antifungal immunity. *Eur J Immunol* 2003, 33:1193-1204.

[75] Chen K, Xu W, Wilson M, He B, Miller NW, Bengtén E, Edholm ES, Santini PA, Rath P, Chiu A, et al.: Immunoglobulin D enhances immune surveillance by activating antimicrobial, proinflammatory and B cell-stimulating programs in basophils. *Nat Immunol* 2009, 10:889-898.

[76] Boita Enrico Heffler Stefano Pizzimenti Alberto Raie Elona Saraci Paola Omedè Claudia Bussolino Caterina Bucca Giovanni Rolla M, Mauriziano Umberto OI: Regulation of B-Cell-Activating Factor Expression on the Basophil Membrane of Allergic Patients. *Int Arch Allergy Immunol* 2015, 166:208-212.

[77] Kim CH: Immune regulation by microbiome metabolites. *Immunology* 2018, 154:220-229.

[78] Galloway-Peña JR, Kontoyiannis DP: The gut mycobiome: The overlooked constituent of clinical outcomes and treatment complications in patients with cancer and other immunosuppressive conditions. *PLOS Pathog* 2020, 16:e1008353.

[79] Wu T, Li H, Su C, Xu F, Yang G, Sun K, Xu M, Lv N, Meng B, Liu Y, et al.: Microbiota-Derived Short-Chain Fatty Acids Promote LAMTOR2-Mediated Immune Responses in Macrophages. *mSystems* 2020, 5.

[80] Schulthess J, Pandey S, Capitani M, Rue-Albrecht KC, Arnold I, Franchini F, Chomka A, Ilott NE, Johnston DGW, Pires E, et al.: The Short Chain Fatty Acid Butyrate Imprints an Antimicrobial Program in Macrophages. *Immunity* 2019, 50:432-445.e7.

[81] Ciarlo E, Heinonen T, Herderschee J, Fenwick C, Mombelli M, Le Roy D, Roger T: Impact of the microbial derived short chain fatty acid propionate on host susceptibility to bacterial and fungal infections in vivo. *Sci Rep* 2016, 6:1-15.

[82] Trompette A, Gollwitzer ES, Yadava K, Sichelstiel AK, Sprenger N, Ngom-Bru C, Blanchard C, Junt T, Nicod LP, Harris NL, et al.: Gut microbiota metabolism of dietary fiber influences allergic airway disease and hematopoiesis. *Nat Med* 2014, 20:159-166.

[83] Ivanov II, Atarashi K, Manel N, Brodie EL, Shima T, Karaoz U, Wei D, Goldfarb KC, Santee CA, Lynch S V., et al.: Induction of Intestinal Th17 Cells by Segmented Filamentous Bacteria. *Cell* 2009, 139:485-498.

[84] McAleer JP, Nguyen NLH, Chen K, Kumar P, Ricks DM, Binnie M, Armentrout RA, Pociask DA, Hein A, Yu A, et al.: Pulmonary Th17 antifungal immunity is regulated by the gut microbiome. *J Immunol* 2016, 197:97-107.

[85] Ganal SC, Sanos SL, Kallfass C, Oberle K, Johner C, Kirschning C, Lienenklaus S, Weiss S, Staeheli P, Aichele P, et al.: Priming of natural killer cells by nonmucosal mononuclear phagocytes requires instructive signals from commensal microbiota. *Immunity* 2012, 37:171-186.

[86] Abt MC, Osborne LC, Monticelli LA, Doering TA, Alenghat T, Sonnenberg GF, Paley MA, Antenus M, Williams KL, Erikson J, et al.: Commensal bacteria calibrate the activation threshold of innate antiviral immunity. *Immunity* 2012, 37: 158-170.

[87] Sonnenberg GF, Artis D: Innate Lymphoid Cell Interactions with Microbiota: Implications for Intestinal Health and Disease. *Immunity* 2012, 37:601-610.

[88] Crellin NK, Trifari S, Kaplan CD, Satoh-Takayama N, Di Santo JP, Spits H: Regulation of cytokine secretion in human CD127+ LTi-like innate lymphoid cells by toll-like receptor 2. *Immunity* 2010, 33:752-764.

[89] Hill DA, Siracusa MC, Abt MC, Kim BS, Kobuley D, Kubo M, Kambayashi T, Larosa DF, Renner ED, Orange JS, et al.: Commensal bacteria-derived signals regulate basophil hematopoiesis and allergic inflammation. *Nat Med* 2012, 18:538-546.

[90] Pragman AA, Kim HB, Reilly CS, Wendt C, Isaacson RE: The Lung Microbiome in Moderate and Severe Chronic Obstructive Pulmonary Disease. *PLoS One* 2012, 7.

[91] Lynch S V.: Viruses and microbiome alterations. In *Annals of the American Thoracic Society*. . Ann Am Thorac Soc; 2014.

[92] Leung RKK, Zhou JW, Guan W, Li SK, Yang ZF, Tsui SKW: Modulation of potential respiratory pathogens by pH1N1 viral infection. *Clin Microbiol Infect* 2013, 19:930-935.

[93] Noverr MC, Noggle RM, Toews GB, Huffnagle GB: Role of antibiotics and fungal microbiota in driving pulmonary allergic responses. *Infect Immun* 2004, 72:4996-5003.

[94] Amin R, Dupuis A, Aaron SD, Ratjen F: The effect of chronic infection with *Aspergillus fumigatus* on lung function and hospitalization in patients with cystic fibrosis. *Chest* 2010, 137:171-176.

[95] Baxter CG, Rautemaa R, Jones AM, Kevin Webb A, Bull M, Mahenthiralingam E, Denning DW: Intravenous antibiotics reduce the presence of *Aspergillus* in adult cystic fibrosis sputum. *Thorax* 2013, 68:652-657.

[96] Kesh S, Mensah NY, Peterlongo P, Jaffe D, Hsu K, VAN DEN Brink M, O'reilly R, Pamer E, Satagopan J, Papanicolaou GA: TLR1 and TLR6 polymorphisms are associated with susceptibility to invasive aspergillosis after allogeneic stem cell transplantation. *Ann N Y Acad Sci* 2005, 1062:95-103.

[97] Lamoth F, Rubino I, Bochud P-Y: Immunogenetics of invasive aspergillosis. *Med Mycol* 2011, 49:S125–S136.

[98] Carvalho A, Pasqualotto AC, Pitzurra L, Romani L, Denning DW, Rodrigues F: Polymorphisms in toll-like receptor genes and susceptibility to pulmonary aspergillosis. *J Infect Dis* 2008, 197:618-621.

[99] de Boer MGJ, Jolink H, Halkes CJM, van der Heiden PLJ, Kremer D, Falkenburg JHF, van de Vosse E, van Dissel JT: Influence of Polymorphisms in Innate Immunity Genes on Susceptibility to Invasive Aspergillosis after Stem Cell Transplantation. *PLoS One* 2011, 6:e18403.

[100] Grube M, Loeffler J, Mezger M, Krüger B, Echtenacher B, Hoffmann P, Edinger M, Einsele H, Andreesen R, Holler E: TLR5 stop codon polymorphism is associated with invasive aspergillosis after allogeneic stem cell transplantation. *Med Mycol* 2013, 51:818-825.

[101] Pamer EG: TLR Polymorphisms and the Risk of Invasive Fungal Infections . *N Engl J Med* 2008, 359:1836-1838.

[102] Cunha C, Di Ianni M, Bozza S, Giovannini G, Zagarella S, Zelante T, D'Angelo C, Pierini A, Pitzurra L, Falzetti F, et al.: Dectin-1 Y238X polymorphism associates with susceptibility to invasive aspergillosis in hematopoietic transplantation through impairment of both recipient- and donor-dependent mechanisms of antifungal immunity. *Blood* 2010, 116:5394-5402.

[103] Chai LYA, de Boer MGJ, van der Velden WJFM, Plantinga TS, van Spriel AB, Jacobs C, Halkes CJM, Vonk AG, Blijlevens NM, van Dissel JT, et al.: The Y238X Stop Codon Polymorphism in the Human β-Glucan Receptor Dectin-1 and Susceptibility to Invasive Aspergillosis. *J Infect Dis* 2011, 203:736-743.

[104] Sainz J, Lupiáñez CB, Segura-Catena J, Vazquez L, Ríos R, Oyonarte S, Hemminki K, Försti A, Jurado M: Dectin-1 and DC-SIGN polymorphisms associated with invasive pulmonary aspergillosis infection. *PLoS One* 2012, 7.

[105] Cunha C, Aversa F, Lacerda JF, Busca A, Kurzai O, Grube M, Löffler J, Maertens JA, Bell AS, Inforzato A, et al.: Genetic PTX3 Deficiency and Aspergillosis in Stem-Cell Transplantation. *N Engl J Med* 2014, 370:421-432.

[106] Wójtowicz A, Lecompte TD, Bibert S, Manuel O, Rüeger S, Berger C, Boggian K, Cusini A, Garzoni C, Hirsch H, et al.: PTX3 Polymorphisms and Invasive Mold Infections after Solid Organ Transplant. *Clin Infect Dis* 2015, 61:619-622.

[107] Sainz J, Pérez E, Hassan L, Moratalla A, Romero A, Collado MD, Jurado M: Variable Number of Tandem Repeats of TNF Receptor Type 2 Promoter as Genetic Biomarker of Susceptibility to Develop Invasive Pulmonary Aspergillosis. *Hum Immunol* 2007, 68:41-50.

[108] Vaid M, Kaur S, Sambatakou H, Madan T, Denning DW, Sarma PU: Distinct alleles of mannose-binding lectin (MBL) and surfactant proteins A (SP-A) in patients with chronic cavitary pulmonary aspergillosis and allergic bronchopulmonary aspergillosis. *Clin Chem Lab Med* 2007, 45:183-186.

[109] Kaur S, Gupta VK, Thiel S, Sarma PU, Madan T: Protective role of mannan-binding lectin in a murine model of invasive pulmonary aspergillosis. *Clin Exp Immunol* 2007, 148:382-389.

[110] Crosdale DJ, Poulton K V., Ollier WE, Thomson W, Denning DW: Mannose-binding lectin gene polymorphisms as a susceptibility factor for chronic necrotizing pulmonary aspergillosis. *J Infect Dis* 2001, 184:653-656.

[111] Borta S, Popetiu R, Donath-Miklos I, Puschita M: Genetic Polymorphism of MBL 2 in Patients

with Allergic Bronchial Asthma. *Maedica (Buchar)* 2019, 14:208-212.

[112] Lambourne J, Agranoff D, Herbrecht R, Buchbinder A, Willis F, Letscher-Bru V, Agrawal S, Doffman S, Johnson E, White PL, et al.: Association of mannose-binding lectin deficiency with acute invasive aspergillosis in immunocompromised patients. *Clin Infect Dis* 2009, 49:1486-1491.

[113] Carvalho A, Cunha C, Di Ianni M, Pitzurra L, Aloisi T, Falzetti F, Carotti A, Bistoni F, Aversa F, Romani L: Prognostic significance of genetic variants in the IL-23/Th17 pathway for the outcome of T cell-depleted allogeneic stem cell transplantation. *Bone Marrow Transplant* 2010, 45:1645-1652.

[114] Zaas AK, Liao G, Chien JW, Weinberg C, Shore D, Giles SS, Marr KA, Usuka J, Burch LH, Perera L, et al.: Plasminogen alleles influence susceptibility to invasive aspergillosis. *PLoS Genet* 2008, 4.

[115] Cunha C, Rodrigues F, Zelante T, Aversa F, Romani L, Carvalho A: Genetic susceptibility to aspergillosis in allogeneic stem-cell transplantation. In *Medical Mycology*. . Oxford Academic; 2011:S137–S143.

[116] Tanpaibule T, Jinawath N, Taweewongsounton A, Niparuck P, Rotjanapan P: Genetic Risk Surveillance for Invasive Aspergillosis in Hematology Patients: A Prospective Observational Study. *Infect Dis Ther* 2020, 9:807-821.

[117] Saxena S, Madan T, Shah A, Muralidhar K, Sarma PU: Association of polymorphisms in the collagen region of SP-A2 with increased levels of total IgE antibodies and eosinophilia in patients with allergic bronchopulmonary aspergillosis. *J Allergy Clin Immunol* 2003, 111:1001-1007.

[118] Madan T, Kaur S, Saxena S, Singh M, Kishore U, Thiel S, Reid KBM, Sarma PU: Role of collectins in innate immunity against aspergillosis. *Med Mycol* 2005, 43:155-163.

[119] Guo Y, Kasahara S, Jhingran A, Tosini NL, Zhai B, Aufiero MA, Mills KAM, Gjonbalaj M, Espinosa V, Rivera A, et al.: During *Aspergillus* Infection, Monocyte-Derived DCs, Neutrophils, and Plasmacytoid DCs Enhance Innate Immune Defense through CXCR3-Dependent Crosstalk. *Cell Host Microbe* 2020, 28:104-116.e4.

[120] Fisher CE, Hohl TM, Fan W, Storer BE, Levine DM, Zhao LP, Martin PJ, Warren EH, Boeckh M, Hansen JA: Validation of single nucleotide polymorphisms in invasive aspergillosis following hematopoietic cell transplantation. *Blood* 2017, 129:2693-2701.

[121] Mezger M, Steffens M, Beyer M, Manger C, Eberle J, Toliat MR, Wienker TF, Ljungman P, Hebart H, Dornbusch HJ, et al.: Polymorphisms in the chemokine (C-X-C motif) ligand 10 are associated with invasive aspergillosis after allogeneic stem-cell transplantation and influence CXCL10 epression in monocyte-derived dendritic cells. *Blood* 2008, 111:534-536.

[122] Lupiañez CB, Martínez-Bueno M, Sánchez-Maldonado JM, Badiola J, Cunha C, Springer J, Lackner M, Segura-Catena J, Canet LM, Alcazar-Fuoli L, et al.: Polymorphisms within the ARNT2 and CX3CR1 genes are associated with the risk of developing invasive aspergillosis. *Infect Immun* 2020, 88:882-901.

[123] Lupiañez CB, Canet LM, Carvalho A, Alcazar-Fuoli L, Springer J, Lackner M, Segura-Catena J, Comino A, Olmedo C, Ríos R, et al.: Polymorphisms in host immunity-modulating genes and risk of invasive aspergillosis: Results

from the AspBIOmics Consortium. *Infect Immun* 2016, 84:643-657.

[124] Brouard J, Knauer N, Boelle PY, Corvol H, Henrion-Caude A, Flamant C, Bremont F, Delaisi B, Duhamel JF, Marguet C, et al.: Influence of interleukin-10 on *Aspergillus fumigatus* infection in patients with cystic fibrosis. *J Infect Dis* 2005, 191:1988-1991.

[125] Seo KW, Kim DH, Sohn SK, Lee NY, Chang HH, Kim SW, Jeon SB, Baek JH, Kim JG, Suh JS, et al.: Protective role of interleukin-10 promoter gene polymorphism in the pathogenesis of invasive pulmonary aspergillosis after allogeneic stem cell transplantation. *Bone Marrow Transplant* 2005, 36:1089-1095.

[126] Sainz J, Hassan L, Perez E, Romero A, Moratalla A, López-Fernández E, Oyonarte S, Jurado M: Interleukin-10 promoter polymorphism as risk factor to develop invasive pulmonary aspergillosis. *Immunol Lett* 2007, 109:76-82.

[127] Knutsen AP, Kariuki B, Consolino JD, Warrier MR: IL-4 alpha chain receptor (IL-4Rα) polymorphisms in allergic bronchopulmonary sspergillosis. *Clin Mol Allergy* 2006, 4.

[128] Lupiañez CB, Villaescusa MT, Carvalho A, Springer J, Lackner M, Sánchez-Maldonado JM, Canet LM, Cunha C, Segura-Catena J, Alcazar-Fuoli L, et al.: Common genetic polymorphisms within NFκB-related genes and the risk of developing invasive aspergillosis. *Front Microbiol* 2016, 7:1243.

Mycotoxin Production and Industrial Application

Chapter 4

The Role of Aflatoxins in *Aspergillus flavus* Resistance to Stress

Massimo Reverberi, Marzia Beccaccioli and Marco Zaccaria

Abstract

Aspergillus section Flavi produce the aflatoxins, secondary metabolites toxic to humans and animals. Why do these fungi produce aflatoxins? They do not have a clear role in pathogenicity or in niche competition. *Aspergillus* employs a considerable amount of energy to synthesize them: more than 20 enzymatic catalyzes are needed. Within the *A. flavus* species, all opportunistic pathogens of maize, more than half of the natural population are atoxigenic, indicating that aflatoxins are not so obviously linked to an enhancement of population fitness. The perspective changes in *A. parasiticus*, pathogen to peanuts, where more than 90% of the natural population produce the four aflatoxins. In this chapter, we aim to discuss our recent hypothesis that aflatoxins act as antioxidants providing more time to *Aspergillus* to "escape" an exploited substrate, that in the meanwhile is "fully charged" with reactive oxygen species and oxylipins.

Keywords: antioxidants, oxylipins, resilience to stress, lifespan, host adaptation

1. Introduction

The species belonging to the section Flavi of the genus *Aspergillus* can produce the carcinogenic aflatoxins (AF), secondary metabolites synthesized as the final product of a very complex pathway including 25 different enzymatic activities so far [1]. Aflatoxins are detrimental for animals (humans included) since, after oxidation by cytochrome P450, their 8,9-epoxide causes DNA depurination leading possibly to carcinogenesis [2]. Notwithstanding these fungi can invade the lungs of cystic fibrosis patients, the role of AF in worsening the clinical frame remains to be demonstrated. *De facto Aspergillus flavus* (and *A. parasiticus* as well) are opportunistic pathogens for animals as well as for plants and a competitive soil saprophyte [3], for which the synthesis of AF appears "luxury" or not necessary. Our idea is that AF are too expensive – in terms of biosynthesis and energy devoted – to be "non necessary". If so, why AF are produced? More than 40 years of research told us that AF are synthesized following different inputs: nutritional [4, 5], pH, light, [6], host defenses [1] and finally, oxidative stress [7, 8]. The general impression is that AF are synthesized in response to a stressful condition. In light of this, can we consider them as a part of the complex reaction set that *Aspergillus* uses for facing challenging conditions? This chapter focus on how oxidative stress can modulate (how and why modulate) AF synthesis, beginning with the description of some of the actors that switch the AF synthesis on; the following paragraph regard the important role of oxidized fatty acids in controlling several aspects of the life of these fungi and, notably, AF synthesis; it concludes with an evolutionary point of view on the meaning of AF synthesis for the *Aspergillus* section Flavi lifestyle.

2. The role of oxidative stress in modulating aflatoxin synthesis

Oxidative stress is a condition which organisms must cope with since the process used for producing energy (namely ATP) involves a very oxidizing molecule: the oxygen [9]. Thus, aerobic organisms have evolved means for facing this stress by building up a complex antioxidant system composed of structures, proteins (enzymes) and small metabolites. The ability to control this system enables organisms to face oxidative stress and, indeed, using it to "boost" some pathways (e.g., the defense in the plants) [10]. Aflatoxins are among these: they are synthesized in response to oxidative stress conditions [7] and, as we aimed to clarify within this chapter, can act as antioxidants to enhance the survival ability of these fungi.

2.1 Reactive oxygen species (ROS)

Free radicals are, by definition, very reactive chemical species, due to their presence of one or more unpaired electrons in valence orbitals. This condition makes them highly reactive molecules, energized and unstable; free radicals will try to to give up or, as more commonly happens, to acquire an electron at the expense of another to obtain a stable configuration.

In living systems, spontaneously forming radicals are numerous, and those of greater biological interest, the so-called ROS, are those molecules in which the unpaired electron is found on O_2, such as, for example, superoxide ($\cdot O_2$-), hydroxyl ($\cdot OH$), hydroperoxyl ($\cdot OOH$), peroxyl ($\cdot OOR$) and alkoxy ($\cdot RO$) radicals. Oxygen is found in nature in the form of diatomic molecules that have two unpaired electrons of parallel spins arranged on two different orbitals (triplet), and therefore possessing characteristics paramagnetic. The fact of having uncoupled electrons makes O_2 particularly prone to forming covalent bonds but, in the case of incomplete reduction, ROS may be generated. These react quickly with other compounds to acquire the electrons necessary for their chemical stability, losing, in turn, their electrons and becoming radicals themselves, thus triggering a chain reaction. Once that process starts, it is determined in the cell a cascade of reactions that often begins with the peroxidation of lipids membrane (oxidation of the hydrocarbon chain), resulting in its destabilization, and which proceeds with the oxidation of other cellular components (such as proteins and DNA), to the point of causing the deconstruction of the entire cell.

The reactions in which the radical molecules can take part are many and vary significantly, for example, depending on: (i) the compartment or organelle cell in which they originate, (ii) of the antioxidant systems present, (iii) of the molecules that they attack, (iv) the water and nutritional conditions of the cell. Also, non-radical molecules, such as hydrogen peroxide (H_2O_2), can trigger responses that lead to the formation of ROS: the Haber-Weiss reaction, for example, produces hydroxyl radicals starting from H_2O_2 and O_2-. The cells of photosynthetic organisms are more subject to oxidative damage since they have concentrations of very high O_2 since, not only do they use it during breathing, but they also generate it with photosynthesis. In fact, they have membrane thylakoids composed mainly of polyunsaturated lipids (molecules subject to reactions of peroxidation) and, by means of photosynthetic pigments, absorb light energy, the excess of which favors the production of ROS. In its ground state, O_2 is relatively not very dangerous because, although it can give rise to excited states reactive and free radicals (during photosynthesis, for example), its utilization proceeds expeditiously by means of a route in stages, in which a reduction to H_2O involving four electrons and during which intermediates can be generated partially reduced reactive species. In fungi as well as in other organisms, ROS can be produced in a tightly regulated way by the NADPH oxidase complex (NOX in fungi; [11]). This complex controls, upon stimuli, the formation of anion superoxide and controls several processes in hyphal growth and development [11, 12] and in mycotoxin synthesis too [13].

2.2 Antioxidant responses

If, as just described, the formation of free radicals can cause serious damage at the cellular level, which can sometimes lead the cell to death, it is equally true that the aerobic cells have evolved and developed efficient ROS control and detoxification systems. The latter are known as antioxidant systems and can be enzymatic and non-enzymatic in nature. The systems non-enzymatic include molecules such as: α-tocopherol, β-carotene, compounds phenolics, ascorbate, glutathione; the enzymatic ones involve: superoxide dismutases (SOD; EC 1.15.1.1), which catalyzes the dismutation of O_2.- in H_2O_2, together to others which eliminate the H_2O_2 such as catalases (CAT; EC 1.11.1.6), peroxidases, glutathione peroxidases (GP; EC 1.11.1.9), (which uses glutathione as an electron donor - GSH, reduced form, and GSSG, oxidized form) and ascorbate peroxidases (APX; EC 1.11.1.11; ASA, reduced form of ascorbic acid, and DHA, oxidized form). All the enzymes described are found in multiple forms (isoforms) that can be classified, for example, based on their metallic cofactor. The latter can also be found in different cellular compartments (such as cytosol and apoplast) or organelles [mitochondrion, chloroplast (in plants), peroxisome and vacuole]. Some of them catalyze the same reaction and can use different substrates as electron donors. In fungi, these antioxidant capacities are tightly controlled by transcriptional regulators. Main transcription factors that in fungi "react" to ROS are msn2–4 [14], skn7 [15] and Yap-1 [16]. Notably, Yap-1 orthologue ApyapA can control aflatoxin biosynthesis [17].

2.3 Oxidative stress, antioxidant system and aflatoxin synthesis

Oxidants are continuously produced within and outside fungal cells. In some way they can fuel cells to switch metabolic pathways [18] or differentiation patterns [19]. Inter alia, in *A. flavus* and *A. parasiticus* ROS boost aflatoxin formation [19–21]. In the past we showed that several oxidants amended to culture as well as increase of cell ROS were able to trigger aflatoxin synthesis [20]. Intriguingly, even external oxidants augment the titer of cell oxidants. How can these oxidants turn into "aflatoxins"? which is the "mediator" of the opening of the complex aflatoxin pathway? Our group and John Linz group demonstrated that ApYapA can orchestrate their synthesis [20]. Notably, ApYapA, similarly to its orthologue Yap-1 of *Saccharomyces cerevisiae*, is indirectly oxidized by ROS through a peroxideroxin (TSA1, [17, 22]. Once oxidized, ApYapA migrates into the nucleus and, during the exponential phase of growth it recognizes and binds to specific responsive elements present into the promoter of genes encoding antioxidant enzymes such as catalases, superoxide dismutases inter alia. During the stationary phase, indeed, it binds even to the promoter of AflR, i.e., the gene whose product controls (together with AflJ) the whole AF pathway and consequently, their biosynthesis (**Figure 1**). As suggested below (paragraph 4), aflatoxins are a subsidiary antioxidant response that fungal cells operate to "staying alive" as long as possible to differentiate conidia and "escape" from the spent - stressing, oxidizing – substrate [23].

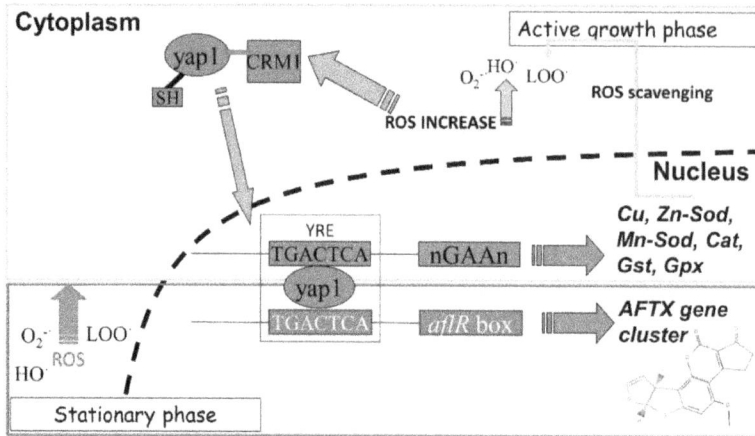

Figure 1.
Aspergillus parasiticus *produces ROS during the normal course of its lifestyle. During the exponential – Active- phase of growth their production is constant but kept low by a very efficient antioxidant system that is modulated by ApyapA inter alia. In this phase aflatoxin synthesis is normally shut off or very low. Indeed, during the stationary phase (or even in consequence of external stressors – E.g., herbicides), ROS scavenging though normal detoxification system (e.g., glutathione, superoxide dismutases etc) is not efficient anymore and the oxidative stress-controlled transcription factor ApYapA recognize and bound to YRE (Yap1 responsive elements) present in the promoter of AflR, the global transcriptional regulator of aflatoxin synthesis. In this phase, AF synthesis is switched on and contribute to scavenge oxidants present in the matrix.*

3. The role of oxylipins in the *Aspergillus* sect. Flavi lifestyle

3.1 Discovery of oxylipins

Understanding the evolution of fungal pathogenesis requires the treatment of some lipid molecules that mediate the fungus-host interaction. *A. flavus* preferentially infects maize seeds, which are rich in unsaturated fatty acids (UFAs). Furthermore, also Aspergillus species contain high levels of UFAs, including oleic ([18]: 1), linoleic ([18]: 2) and linolenic ([18]: 3) acid, which are substrates for oxygenation that converts the UFAs in oxylipins. Oxylipins are involved in the interaction-signaling of fungi, bacteria, plants and animals.

First evidence on the existence of oxylipins dates to 1987, when Champe et al., demonstrate the role of precocious sexual inducer (psi), later called oxylipins, in Aspergillus nidulans. Psi factors inhibit asexual sporulation and stimulate premature sexual sporulation, acting as hormone-like molecule [24].

Oxylipins derive from free fatty acids or from fatty acids present into membrane phospholipids. Fatty acids included in membranes, during the plant-pathogen interaction, are released by lipase action. Lipases are considered as virulence factors in plant pathogenic fungi [25].

Oxylipin oxidation may happen by two routes: the radical and the enzymatic. In fact, during the first steps of infection the production of radical species favors the accumulation of Reactive Oxygen Species (ROS). Superoxide anion ($\cdot O_2$-), hydrogen peroxide (H_2O_2) and hydroxyl radical ($\cdot OH$) can spontaneously oxidize the free fatty acids. The second oxylipins synthesis route, i.e., enzymatic, in fungi involves the action of dioxygenases (DOX), lipoxygenases (LOX) and cyclooxygenases (COX) that convert free fatty acids in oxylipins [26].

The crosstalk established during a plant-pathogenic fungus interaction, therefore, involves a lipases-LOX concerted activity, that carries towards the oxylipin biogenesis [27].

In *A. flavus* enzymatic set for oxylipin formation is composed of four dioxygenases, PpoA, PpoB, PpoC, and PpoD, and one lipoxygenase, LOX. PpoA encodes a 5,8-linoleate diol synthase, whereas ppoC encodes for a linoleate (10R)-dioxygenase [28–30].

3.2 Oxylipins in host-pathogen interaction

Oxylipins act as modulator of many signal transduction pathways, both in plant and fungi because the chemical structure as well as the main synthesis routes of oxylipins are common between the two kingdoms. For that reason, several authors defined the oxylipins as the common language between hosts and pathogens [31]. But why do hosts and pathogens produce oxylipins?

In *A. flavus*, oxylipins regulate dissemination, affecting the production of sclerotia and conidia, and influence secondary metabolism. In addition to having an autocrine action, oxylipins are also distinguished by their paracrine action, when *A. flavus* interacts with other organisms.

When *A. flavus* produces oxylipins, the plant recognized them and alters the expression of oxylipin synthesis genes, as the LOX, but the fungus senses the plant oxylipin gradient that promotes sporulation and mycotoxin production, this signaling exchanging define the cross-talk.

Plant can release oxylipins as 9S-HPODE (9S-hydroperoxyoctadecadienoic acid) and 13S-HPODE (13S-hydroperoxyoctadecadienoic acid) able to influence the development in the Aspergilli, in addition, but at the same time they act in the regulation of plant defense and development.

The analysis of the phenotypes derived from the mutant of ppo-genes shown that in *A. flavus* that ppoA and ppoC deletion generates strains with less conidia and more sclerotia, whereas the deletion of ppoD shown the inverted situation, or rather more conidia and less sclerotia. It was considered also the deletion mutant for all four oxylipin-biosynthesis genes (dioxygenases and lox), which shows both high levels of aflatoxin production and high levels of sclerotia production. These results shown the closely link between the oxylipin production and the asexual or sexual reproduction, underlining the role of the oxylipins in the fungal regeneration.

As previously introduced, the oxylipins produced by the plant may influence the fungal lifestyle and being chemically similar to the fungal oxylipins they can substitute them. That was demonstrated in one study, where the maize lipoxygenase Zm-LOX3 cloned in *A. nidulans* mutant strain, deficient for the two genes ppoA and ppoC, restored them functions. Oxylipins for the plant assume a protective function, in fact the inactivation of Zm-LOX3 makes the plant more susceptible to *A. flavus*, that during the infection grows more and produces more aflatoxins [32]. The increase of susceptibility is linked to the accumulation of oxylipin substrates, the fatty acids, and the decrease of the jasmonic acid, in whose biosynthesis pathway Zm-LOX3 is involved. Oxylipins acting as hormones, that means that a small concentration of oxylipins can regulate physiological processes as growth and development, but several studies show that oxylipin-mediated responses are strongly influenced by the type and the nature of interaction with the host.

The development of the fungus depends on cell densities when the cell density is high *A. flavus* produces more conidia. The cell densities also influence the secondary metabolites production, in fact the aflatoxin synthesis decreases at high cell densities. The quorum sensing, or rather the phenomenon in which the set of signaling molecules enable the single cell to sense the other cells, may be associated with the oxylipins release. The deletion of ppo and lox genes inhibits the development of the oxylipins [33, 34]. The exposure to exogenous seed oxylipins, as 9-HpODE and 13-HpODE stimulate the sporulation and the aflatoxin synthesis. The 13-HpODE

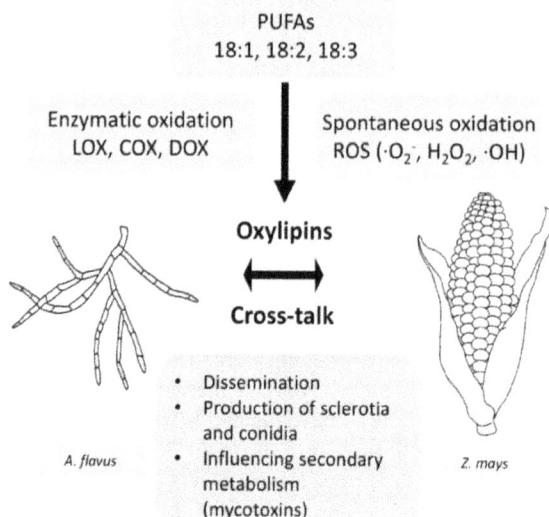

Figure 2.
Signaling oxylipin-mediated in A. flavus. *Polyunsaturated fatty acids (PUFAs) may be converted by enzymatic and spontaneous oxidation in oxylipins, that mediate the interaction with the host (Z. mays).*

seems to inhibit the sclerotia formation, suggesting the role of this oxylipin in the sexual/asexual reproduction in *Aspergillus* species [35]. These results proposed the oxylipins as quorum sensing mediator in *Aspergillus* species [36] (**Figure 2**).

Although the role of oxylipins in development and host-pathogen communication is recognized, little is known about their perception. Mammal oxylipins are sensing by G protein-coupled receptors (GPCRs), but in fungi GPCRs are not actually identified also if in *A. flavus* have been found some genes homologous to mammal GPCRs required for the high-density growth [37].

GPCR-mediated signaling seems to be linked to pathogenesis, therefore it is hypothesized that they could be potential targets for disease control [38].

4. Evolving a strategy for enhancing the resilience to stress: the role of aflatoxins

A. flavus is highly infectant, fast growing, efficient and versatile bio factory to its core [39]. Among the wide inventory of secondary metabolites it produces, aflatoxins are not the least puzzling, even after decades of dedicated research. Their physiological role has proven hard to frame unequivocally, having been linked over the years to several different purposes, including messengers for quorum sensing, facilitators of dispersal and resistance factors to UV stress [40], inhibitors of environmental competitors [41], mutagenic agents employed to compensate the limited intraspecific variability derived by a conidiogenesis-centred reproduction strategy and, finally, antioxidants.

Aflatoxins have been linked to oxidative stress since the 1980s [42], their synthesis being convincingly linked to increase in the presence of oxidants, *in vitro* and *in vivo* [43, 44]. However, the mechanism behind aflatoxins' role as direct or indirect reducing agents has also been elusive. In a general view, secondary metabolites such as mycotoxins are described by several experts as the by-product of housekeeping processes whose production is an indirect consequence of the very activity of primary metabolism [45]. In this framework, it is possible that aflatoxins could fit the picture as the by-products of a secondary pathway aimed at channeling and exhausting environmental oxidative stress through a dedicated metabolic pathway, which only culminates in aflatoxin synthesis; aflatoxins would therefore achieve their alleged biological purpose through an indirect role. This eventuality is intriguing but, also, does not preclude the avenue of a more direct role as a ROS scavenger. As all secondary metabolites, aflatoxin production is mainly triggered after the biological switch from trophophase to idiophase, when *Aspergillus* transitions from a growth-centred lifestyle to differentiation and dispersal. At such a stage of mycelial development, aflatoxin must provide a benefit to growth/survivability to an appreciable, but not substantial extent. Given that even within *A. flavus* strains only half are aflatoxigenic, it would be sensible to hypothesize that aflatoxins in no way define or drive *A. flavus* ecology, but also in no way do they burden it beyond redemption, otherwise they would be excised from secondary metabolism altogether through selection, if nothing else, in the context of evolution.

A. parasiticus is an aflatoxin producer, and close relative to *A. flavus,* with a well-documented tolerance to intense oxidative stress, both *in vitro* and *in vivo*. Hong et al. [46] have provided precious insight into *A. parasiticus* aflatoxigenic biological triggers. In their work, antioxidant enzymes are upregulated during growth stage and into early stationary phase; it is only once stably into stationary phase that the oxidative stress-related transcription factory AP-1 like triggers aflatoxin synthesis. In this context, it is easy to see how strongly an intrinsic, direct antioxidant potential of the aflatoxin molecule would constitute a substantial clue to the proof of mechanism researchers have wondered about.

A recent research by Finotti et al. [47] aims at elucidating this aspect. Finotti explored the intrinsic potential (**Figure 3**) of the four main AFs congeners (B1, B2, G1, G2) as scavengers of reactive oxygen species (ROS). In this work, 2,2′-Azobis,

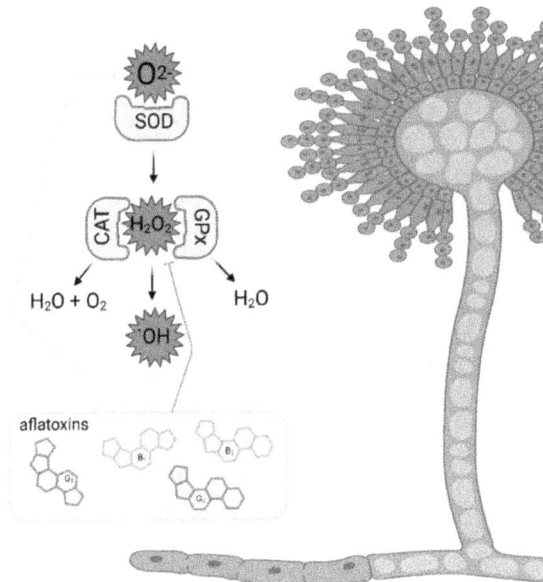

Figure 3.
ROS are highlighted with a red star. Antioxidant enzymes superoxide dismutase (SOD), catalase (CAT), and glutathione peroxidase (GPx) are adjuvated by aflatoxins' scavenging activity on H_2O_2 (observed), and on O^{2-} and ·OH (putative). Created with BioRender.com.

2-amidinopropane (APAB) was used to generate oxidants *in vitro* in hydrophilic and lipophilic environments. In the former case, all aflatoxin variants proved capable at inhibiting the oxidant-induced bleaching of crocin, each with different degrees of efficacy, namely: G1 > B2 > G2 > B1, ranked from most to least effective. Notably, AFG1 presented an antioxidant value (Ka/Kc = 2.49) comparable to that of the hydrophilic fraction of select polyphenols known for their remarkable antioxidant activity. A second *in vitro* test was run by Finotti and collaborators to assess survivability of *E. coli* K12 cells, when faced with hydrogen peroxide-induced oxidative stress, in the presence of aflatoxin B1. *E. coli* was selected for the test because it is non-susceptible to aflatoxin toxicity, most likely due to its lack of cytochrome p450, the enzyme whose interaction is necessary to incur into the toxic effect of AFs. 20 ug/mL of AFB1 provided increased carrying capacity to populations of *E.coli* K12 when challenged with hydrogen peroxide concentrations within 0 and 0.6 mM, and no difference with the control beyond such interval. It is of note how Finotti's data substantiate the hypothesis of AFs as active scavengers of ROS. We do not think, however, that this evidence disproves in any way the considerations on AFs' putative, ulterior biological roles. It is indeed a complicated endeavor to frame the purpose of a molecule whose ecological role is likely residual, and whose benefit to the producing organism arguably pertains more than one aspect of life, but possibly none decisively.

5. Conclusions

Fungal species belonging to the *Aspergillus* sect. Flavi synthesize the animal health-hazardous aflatoxins, the most potent natural carcinogenic on earth. Why do these fungi produce them? Our suggestion, recently supported by our findings and by other scientists, is that aflatoxins are a way for resisting an oxidizing environment; namely, they are produced to provide fungi "more time" to induce conidiogenesis and - literally - escape from the stressing environment. Considering this, oxidant stressors produced in different contexts - ranging from herbicide treated soil to host tissues - trigger aflatoxin biosynthesis that - in turn - enhance the antioxidant capacity of *Aspergillus* sect. Flavi and provide a chance to face challenging environments and exploit them.

Acknowledgements

Several young and "less" young scientists have a minor but important role in this play call "why Aspergillus produces aflatoxins": I reserve a special thanks to Corrado Fanelli, Anna Adele Fabbri, Enrico Finotti, Alessandra Ricelli, Marzia Scarpari, Marta Punelli and Valeria Scala *inter alia*.

Conflict of interest

"The authors declare no conflict of interest."

The Role of Aflatoxins in Aspergillus flavus *Resistance to Stress*
DOI: http://dx.doi.org/10.5772/intechopen.99003

Author details

Massimo Reverberi[1*], Marzia Beccaccioli[1] and Marco Zaccaria[2]

1 Department of Environmental Biology, Sapienza University, Roma, Italy

2 Department of Biology, Boston College, Chestnut Hill, MA, United States

*Address all correspondence to: massimo.reverberi@uniroma1.it

IntechOpen

References

[1] Yu J. 2012. Current understanding on aflatoxin biosynthesis and future perspective in reducing aflatoxin contamination. Toxins (Basel) 4:1024-1057.

[2] Benkerroum N. 2020. Chronic and acute toxicities of aflatoxins: Mechanisms of action. Int J Environ Res Public Health 17:423.

[3] Frisvad JC, Hubka V, Ezekiel CN, Hong S-B, Nováková A, Chen AJ, Arzanlou M, Larsen TO, Sklenář F, Mahakarnchanakul W. 2019. Taxonomy of aspergillus section Flavi and their production of aflatoxins, ochratoxins and other mycotoxins. Stud Mycol 93:1-63.

[4] Ma H, Zhang N, Sun L, Qi D. 2015. Effects of different substrates and oils on aflatoxin B 1 production by aspergillus parasiticus. Eur Food Res Technol 240:627-634.

[5] Liu J, Sun L, Zhang N, Zhang J, Guo J, Li C, Rajput SA, Qi D. 2016. Effects of nutrients in substrates of different grains on aflatoxin B1 production by aspergillus flavus. Biomed Res Int 2016.

[6] Caceres I, Khoury A Al, Khoury R El, Lorber S, Oswald IP, Khoury A El, Atoui A, Puel O, Bailly J-D. 2020. Aflatoxin biosynthesis and genetic regulation: A review. Toxins (Basel) 12:150.

[7] Jayashree T, Subramanyam C. 2000. Oxidative stress as a prerequisite for aflatoxin production by aspergillus parasiticus. Free Radic Biol Med 29:981-985.

[8] Goltapeh EM, Aggarwal R, Pakdaman BS. Renu. (2007): Molecular characterization of aspergillus species through amplicon length polymorphism (ALP) using universal rice primers. J Agric Sci Technol 3:29-37.

[9] Hõrak P, Cohen A. 2010. How to measure oxidative stress in an ecological context: Methodological and statistical issues. Funct Ecol 24:960-970.

[10] Noctor G, Reichheld J-P, Foyer CH. 2018. ROS-related redox regulation and signaling in plants, p. 3-12. *In* Seminars in Cell & Developmental Biology. Elsevier.

[11] Ryder LS, Dagdas YF, Mentlak TA, Kershaw MJ, Thornton CR, Schuster M, Chen J, Wang Z, Talbot NJ. 2013. NADPH oxidases regulate septin-mediated cytoskeletal remodeling during plant infection by the rice blast fungus. Proc Natl Acad Sci 110:3179-3184.

[12] Nordzieke DE, Fernandes TR, El Ghalid M, Turrà D, Di Pietro A. 2019. NADPH oxidase regulates chemotropic growth of the fungal pathogen fusarium oxysporum towards the host plant. New Phytol 224:1600-1612.

[13] Reverberi M, Punelli M, Scala V, Scarpari M, Uva P, Mentzen WI, Dolezal AL, Woloshuk C, Pinzari F, Fabbri AA. 2013. Genotypic and phenotypic versatility of aspergillus flavus during maize exploitation. PLoS One 8.

[14] Hasan R, Leroy C, Isnard A, Labarre J, Boy-Marcotte E, Toledano MB. 2002. The control of the yeast H_2O_2 response by the Msn2/4 transcription factors. Mol Microbiol 45:233-241.

[15] Basso V, Znaidi S, Lagage V, Cabral V, Schoenherr F, LeibundGut-Landmann S, d'Enfert C, Bachellier-Bassi S. 2017. The two-component response regulator Skn7 belongs to a network of transcription factors regulating morphogenesis in Candida albicans and independently limits morphogenesis-induced ROS

accumulation. Mol Microbiol 106:157-182.

[16] Mendoza-Martínez AE, Cano-Domínguez N, Aguirre J. 2020. Yap1 homologs mediate more than the redox regulation of the antioxidant response in filamentous fungi. Fungal Biol 124:253-262.

[17] Reverberi M, Zjalic S, Ricelli A, Punelli F, Camera E, Fabbri C, Picardo M, Fanelli C, Fabbri AA. 2008. Modulation of antioxidant defense in aspergillus parasiticus is involved in aflatoxin biosynthesis: A role for the ApyapA gene. Eukaryot Cell 7:988.

[18] Montibus M, Pinson-Gadais L, Richard-Forget F, Barreau C, Ponts N. 2015. Coupling of transcriptional response to oxidative stress and secondary metabolism regulation in filamentous fungi. Crit Rev Microbiol 41:295-308.

[19] Tudzynski P, Heller J, Siegmund U. 2012. Reactive oxygen species generation in fungal development and pathogenesis. Curr Opin Microbiol 15:653-659.

[20] Zaccaria M, Ludovici M, Sanzani SM, Ippolito A, Cigliano RA, Sanseverino W, Scarpari M, Scala V, Fanelli C, Reverberi M. 2015. Menadione-induced oxidative stress re-shapes the oxylipin profile of aspergillus flavus and its lifestyle. Toxins (Basel) 7:4315-4329.

[21] Fabbri A A, Fanelli C, Panfili G, Passi S, Fasella P. 1983. Lipoperoxidation and aflatoxin biosynthesis by Aspergillus parasiticus and Aspergillus flavus.

[22] Roze L V, Chanda A, Wee J, Awad D, Linz JE. 2011. Stress-related transcription factor AtfB integrates secondary metabolism with oxidative stress response in aspergilli. J Biol Chem 286:35137-35148.

[23] Hong S, Roze L V, Wee J, Linz JE. 2013. Evidence that a transcription factor regulatory network coordinates oxidative stress response and secondary metabolism in aspergilli. Microbiologyopen 2:144-160.

[24] Champe SP, Rao P, Chang A. 1987. An endogenous inducer of sexual development in aspergillus nidulans. Microbiology 133:1383-1387.

[25] Gaillardin C. 2010. Lipases as Pathogenicity Factors of Fungi. Handb Hydrocarb lipid Microbiol.

[26] Beccaccioli M, Reverberi M, Scala V. 2019. Fungal lipids: Biosynthesis and signalling during plant-pathogen interaction. Front Biosci - Landmark https://doi.org/10.2741/4712.

[27] Bonaventure G. 2014. Lipases and the biosynthesis of free oxylipins in plants. Plant Signal Behav 9:e28429.

[28] Brodhun F, Göbel C, Hornung E, Feussner I. 2009. Identification of PpoA from aspergillus nidulans as a fusion protein of a fatty acid heme dioxygenase/peroxidase and a cytochrome P450. J Biol Chem 284:11792-11805.

[29] Brodhun F, Schneider S, Göbel C, Hornung E, Feussner I. 2010. PpoC from aspergillus nidulans is a fusion protein with only one active haem. Biochem J 425:553-565.

[30] Garscha U, Jernerén F, Chung D, Keller NP, Hamberg M, Oliw EH. 2007. Identification of dioxygenases required for aspergillus development. J Biol Chem 282:34707-34718.

[31] Christensen SA, Kolomiets M V. 2011. The lipid language of plant-fungal interactions. Fungal Genet Biol.

[32] Gao X, Brodhagen M, Isakeit T, Brown SH, Göbel C, Betran J, Feussner I, Keller NP, Kolomiets M V.

2009. Inactivation of the lipoxygenase ZmLOX3 increases susceptibility of maize to aspergillus spp. Mol plant-microbe Interact 22:222-231.

[33] Brown SH, Zarnowski R, Sharpee WC, Keller NP. 2008. Morphological transitions governed by density dependence and lipoxygenase activity in aspergillus flavus. Appl Environ Microbiol 74:5674-5685.

[34] Brown SH, Scott JB, Bhaheetharan J, Sharpee WC, Milde L, Wilson RA, Keller NP. 2009. Oxygenase coordination is required for morphological transition and the host–fungus interaction of *aspergillus flavus*. Mol Plant-Microbe Interact 22:882-894.

[35] Calvo AM, Hinze LL, Gardner HW, Keller NP. 1999. Sporogenic effect of polyunsaturated fatty acids on development of aspergillus spp. Appl Environ Microbiol 65:3668-3673.

[36] Amaike S, Keller NP. 2011. Aspergillus flavus. Annu Rev Phytopathol 49:107-133.

[37] Affeldt KJ, Brodhagen M, Keller NP. 2012. Aspergillus oxylipin signaling and quorum sensing pathways depend on G protein-coupled receptors. Toxins (Basel) 4:695-717.

[38] Brown NA, Schrevens S, Van Dijck P, Goldman GH. 2018. Fungal G-protein-coupled receptors: Mediators of pathogenesis and targets for disease control. Nat Microbiol 3:402-414.

[39] Moore GG. 2021. Practical considerations will ensure the continued success of pre-harvest biocontrol using non-aflatoxigenic aspergillus flavus strains. Crit Rev Food Sci Nutr 1-18.

[40] Erlich KC. 2006. Evolution of the aflatoxin gene cluster. Mycotoxin Research. 22 (1) 9-15

[41] Drott MT, Lazzaro BP, Brown DL, Carbone I, Milgroom MG. 2017. Balancing selection for aflatoxin in aspergillus flavus is maintained through interference competition with, and fungivory by insects. Proc R Soc B Biol Sci 284:20172408.

[42] Fanelli C, Fabbri A. A, Finotti E, Fasella P, Passi, S.1984. Free radicals and aflatoxin biosynthesis. Experientia 40, 191-193.

[43] Paciolla C, Dipierro N, Mulè G, Logrieco A, Dipierro S. 2004. The mycotoxins beauvericin and T2 induce cell death and the alteration to the ascorbate metabolism in tomato protoplast. Physiol. Mol. Plant Pathol. 65, 49-56.

[44] Reverberi M, Ricelli A, Zjalic, Fabbri AA, Fanelli C. 2010. Natural functions of mycotoxins and control of their biosynthesis in fungi. Appl. Microbiol. Biotechnol. 87, 899-911.

[45] Roze LV, Chanda A, Linz JE. 2010. Compartmentalization and molecular traffic in secondary metabolism: A new understanding of established cellular processes. Fungal Genetics and Biology 48 (1) 35-48.

[46] Hong SY, Roze LV, Linz JE. 2013. Oxidative stress-related transcription factors in the regulation of secondary metabolism. Toxins. 5, 683-702.

[47] Finotti E, Parroni A, Zaccaria M, Domin M, Momeni B, Fanelli C, Reverberi M. 2021. Aflatoxins are unorthodox scavengers of reactive oxygen species. Scientific Reports, *accepted for publication*.

Chapter 5

Mycovirus Containing *Aspergillus flavus* and Acute Lymphoblastic Leukemia: Carcinogenesis beyond Mycotoxin Production

Cameron K. Tebbi, Ioly Kotta-Loizou and Robert H.A. Coutts

Abstract

Carcinogenic effects of *Aspergillus* spp. have been well established and generally attributed to a variety of mycotoxin productions, particularly aflatoxins. It is known that most carcinogenic mycotoxins, with the exception of fumonisins, are genotoxic and mutagenic, causing chromosomal aberrations, micronuclei, DNA single-strand breaks, sister chromatid exchange, unscheduled DNA synthesis *etc*. Some *Aspergillus* spp. are infected with mycoviruses which can result in loss of aflatoxin production. The effects of mycovirus containing *Aspergillus* on human health have not been fully evaluated. Recent studies in patients with acute lymphoblastic leukemia, in full remission, have revealed the existence of antibody to the products of a certain *Aspergillus flavus* isolate which harbored an unknown mycovirus. Exposure of blood mononuclear cells from these patients, but not controls, to the products of this organism had reproduced cell surface phenotypes and genetic markers, characteristic of acute lymphoblastic leukemia. Carcinogenic effects of *Aspergillus* spp. may not always be mycotoxin related and this requires further investigation.

Keywords: Acute lymphoblastic leukemia, Mycovirus, Aspergillus, Cancer, Etiology, Leukemogenesis, Carcinogenesis, Virus, Mycotoxin

1. Introduction

With a worldwide distribution and a significant level of genetic diversity, fungi are of importance in both medical and agricultural fields and represent major health and commercial concerns. Medically, fungal organisms can be a part of the normal flora of humans and animals. However, these also have the potential to cause mild to severe life-threatening invasive infections or toxicities. The immune response to fungal agents is variable and complex, ranging from lack of recognition to severe inflammatory reactions resulting in significant morbidity and mortality [1–6].

There is a broad and diverse spectrum of human and animal diseases attributed to fungi. Major effects of fungal agents in human health include, but are not limited to, organ-specific and systemic infections, especially in immunocompromised individuals, toxicity emanating from fungal products, carcinogenicity, mutagenicity, growth impairment and stimulation of allergic reactions. Common and usually non-life-threatening infections caused by fungal agents affecting humans are well recognized and often localized on nails, skin, oral cavity, throat and vagina. Severe and fatal infections, however, can be caused by a variety of fungi including *Aspergillus, Blastomyces, Candida, Coccidioides, Cryptococcus, Histoplasma, Mucoromycetes, Pneumocystis, Talaromyces, etc.* Despite the significance of fungal infections an understanding of their pathophysiology has lagged behind other human pathogens. While the immune system of healthy individuals, in general, can effectively prevent some fungal infections, this is not the case in immunosuppressed patients [7, 8].

In addition to causing direct infections, the products of some fungal organisms can be toxic to animals and humans. Also, the mycobiome has been implicated in the pathogenesis of various types of cancers. An example is the link between *Malassezia spp.* and development of pancreatic ductal adenocarcinoma (PDA) [9]. Based on a reported murine experiment, fungal migration from the intestinal lumen to the pancreas initiates the pathogenesis of PDA by driving the complement cascade through the activation of mannose-binding lectin (MBL) [10]. Another example is the carcinogenic potential of *Candida spp.* Some findings indicate that *Candida albicans* is capable of promoting cancer by several mechanisms, including production of carcinogenic byproducts, inflammation, induction of T helper type 17 (Th17) cell response and molecular imitations [10–12]. As will be discussed later in this article, possible relationships between fungal agents and hematological malignancies have been explored.

In light of the above, here the well-established significance of mycotoxins in carcinogenesis is discussed and novel findings illustrating that mycovirus infections may also play a role in human diseases is highlighted. In particular, focus is placed on a mycovirus containing *Aspergillus flavus* and its effects on leukemogenesis.

2. Mycotoxins

The toxicity, mutagenic and carcinogenic effects of some fungi is often attributed to their production of mycotoxins. Mycotoxins are low molecular weight metabolites produced by yeasts and filamentous fungi. These metabolites are heterogeneous chemicals, toxic to vertebrates, including humans. Several mycotoxins also have toxicities to invertebrates, plants, and other microorganisms [13, 14].

Currently, there are over 450 known mycotoxins, which along with their secondary metabolites, can produce varying degrees of toxicity ranging from mild gastrointestinal symptoms to cancer. A large number of common mycotoxins have been identified that are of major concern to human health, among which are aflatoxins, fumonisins, ochratoxins, patulin, zearalenone and nivalenol/deoxynivalenol. Some organisms can produce several different mycotoxins, and many different species may produce the same mycotoxins. Mycotoxin producing fungi are usually found in improperly saved edibles and agricultural commodities. They can enter and contaminate human and animal food supplies. Animals fed contaminated foods can pass aflatoxins through their eggs, milk, and meats, thus indirectly transmitting aflatoxins to humans [15, 16]. While toxicity in humans is often due to ingestion of large doses of mycotoxins, these can also permeate through the skin [17].

Many mycotoxins are cytotoxic and suppress the functions of lymphocytes, granulocytes, and monocytes. Exposure to some mycotoxins inhibits interferon gamma producing Th1 cells and results in decreased number of these cells. Mycotoxins may lead to T cell polarization toward the Th2 phenotype and is a risk factor for the development of allergies [18–23]. The principal function of Th1 cells is cell-mediated immunity and inflammation. In normal conditions, there is a balance between Th1 and Th2 cells. A shift of such a balance results in various disorders. Th1 cells play an important role in the functions of immunity related cells such as macrophages, B cells, and cytotoxic CD8$^+$ T lymphocytes (CTLs). The latter stimulate cellular immune response, participate in the inhibition of the activation of macrophages and invigorate B cells to produce IgM and IgG1. For instance, it is found that T cells of children exposed to *Aspergillus* have significantly lower Th1 cytokines, including tumor necrosis factors (TNFs), interferon-γ, interleukin-2 and -10. These cytokines are involved in the development of CTLs and natural killer (NK) cells which are responsible for the cell-mediated immune response against viruses and detection and removal of tumor cells. Thus, exposure to fungal agents may significantly change cellular composition and cytokine production and immune function [24, 25].

Exposure to aflatoxins can lead to life threatening acute poisoning (aflatoxicosis) [26]. In turn, acute aflatoxicosis can result in acute hepatic necrosis often manifested by symptoms of liver failure [27]. This eventually may cause development of cirrhosis in the liver and hepatic carcinoma. Chronic low-level exposure to mycotoxins, particularly aflatoxins and especially aflatoxin B1, is known to be associated with increased risk of hepatic damage, liver and gallbladder cancer and impaired immune activity [27–29]. Several studies have documented liver and gallbladder toxicity and carcinogenicity related to mycotoxins. Other organs, including bones, kidneys, pancreas, bladder, viscera and central nervous system, can be subject to carcinogenesis [30].

A variety of mycotoxins have carcinogenic potential in animals and humans [16, 17, 26, 28, 31–35]. Certain mycotoxins, especially aflatoxins, produced by genetically diverse *Aspergillus* spp. including *A. fumigatus, A. parasiticus* and *A. flavus* can be genotoxic with damage to DNA, which is attributed to the development of cancer in animals and humans. The effects of aflatoxins B1, B2, G1 and G2 and their metabolites such as aflatoxins M1, M2a, P1, Q1, Q2a, R0, H1; B2a, M2; GM1, GM2a, parasiticol (B3) and GM2, produced by the *Aspergillus* spp., are well recognized [35].

The carcinogenesis of mycotoxins is reported to be due to the intercalation of aflatoxin metabolites into DNA which alkylate the bases through epoxide moiety. This can be as a result of the mutations in the *p53* gene or signaling apoptosis. The third base of codon 249 of the *p53* gene is reported to be more susceptible to aflatoxin-mediated mutations. For example, in hepatocellular carcinoma, upon exposure to aflatoxin, mutation of *p53* gene is fixed at codon 249 third base and take the form of G to T transversion [36, 37].

In one report, using a mammalian cell line, the mutagenicity of various mycotoxins and the efficiency of mutagenic mycotoxins in producing DNA single strand breaks and chromosome aberrations were investigated. These experiments revealed that aflatoxin B1, mycophenolic acid, patulin, penicillic acid, and sterigmatocystin induce 8-azaguanine-resistant mutations. At higher concentrations, aflatoxin B1, mycophenolic acid, and sterigmatocystin were found to have minimal effects on single-stranded DNA. In contrast, treatment with patulin and penicillic acid at higher concentrations had resulted in severe breaks. Chaetoglobosin B, fusarenon X, luteoskyrin, and ochratoxin A had not induced 8-azaguanine-resistant mutations [38].

Overall, the mutagenicity of mycotoxins varies significantly and depends on their efficiency in causing DNA single-strand breaks, resulting in chromosomal aberrations. Adults are believed to have a higher tolerance to mycotoxins but exposure of children, while controversial and not uniformly accepted, can lead to delayed development and stunted growth [16, 31–33].

In addition to laboratory-based experiments, reports regarding isolation of mycotoxin producing strains of fungi, including that of *A. flavus*, from the residences of leukemia patients are available [39–42]. In many reports, except for recent publications, fungal carcinogenesis is attributed to mycotoxins and their immunosuppressive effects. One report describes examination of sera from 36 cancer patients against an aflatoxin producing *A. flavus* which was isolated from the home of a patient with leukemia. A modified microimmunodiffiusion technique was used for this immunological evaluation. This study had found that 30% of cancer patients, 15 of whom had leukemia or lymphoid malignancy, and 6% of controls had shown a precipitation band indicating positive results [39]. Another published article reports four leukemic patients, from three families, in a residence where a mycotoxin producing fungus was isolated. The leukemogenesis was attributed to the immune depressive effects of mycotoxins [41]. In a house where a husband and wife had developed acute myelomonocytic and undifferentiated leukemia, respectively, fungal surveyance of the residence had been performed. Three fungal isolates were found, an extract of which had shown a depressive effect on a phytohemagglutinin skin test in guinea pigs as compared to negative findings using extracts isolated from a control residence [40]. As described below, a significant amount of data regarding the correlation of a mycovirus containing *A. flavus*, isolated from the home of a patient with acute lymphoblastic leukemia, has been recently published.

3. Viruses and human cancer

A vast amount of data on several viruses and their possible association with cancer development has been published [43–52]. While not the focus of this article, a brief review of the subject reveals the importance of the study of viral agents and their relation to occurrence of malignant disorders. Both DNA and RNA viruses are capable of causing cancer in humans. Some of the known DNA viruses that are capable of causing human cancers are Epstein-Barr (EB) virus, human papilloma virus, hepatitis B virus, and human herpes virus 8. The relationship of EB virus to the development of Burkitt's lymphoma and nasopharyngeal carcinoma is well established [53–59]. Likewise, the relation of human papilloma virus and the development of cervical cancer and retention of HPV viral oncoproteins E6 and E7 for their continued expression and proliferation has been demonstrated [60–63]. Human T lymphotropic virus type 1, human immunodeficiency virus (HIV) and hepatitis C viruse are some of the RNA viruses that contribute to human cancers. It appears that viruses have diverse biological pathways to malignant disorders. The presence of viral gene products in cancer and precancerous cells are known. Despite the well-known carcinogenic role of viruses, little data regarding any possible health effects of mycoviruses alone, or in conjunction with their host, are available. This area needs to be further explored.

4. Mycoviruses

Viruses that infect fungi, also known as mycoviruses (*myco* = 'fungus' in Greek), are widespread geographically and are expected to infect all fungal taxa, from early divergent lineages to the most well-studied ascomycetes (sac fungi) and basidiomycetes (mushrooms). Mycovirus infection is persistent but does not result in disease or death of the host fungus, and often does not lead to obvious alterations in its phenotype under controlled laboratory conditions; therefore, mycovirology is an underappreciated and understudied field, similar to all non-disease associated virology [64].

Mycoviruses are currently classified in 22 taxa (21 families and one genus) by the International Committee on Taxonomy of Viruses (ICTV; https://talk.ictvon-line.org/) (**Figure 1**). Some of these taxa exclusively accommodate viruses infecting fungi, such as the families *Hypoviridae* and *Polymycoviridae*. Other taxa also accommodate viruses infecting protozoa, plants, insects and mammals, such as the families *Botourmiaviridae*, *Chrysoviridae*, *Partitiviridae*, *Reoviridae* and *Totiviridae*. Members of the DNA-containing *Genomoviridae* family have been discovered in sequencing data from a variety of samples, including plant and insect tissue, animal blood, serum and feces, human blood, plasma, cerebrospinal fluid, cervical biopsies, and feces, and sewage [65]. Mycoviruses may be closely related to viruses pathogenic for humans. For instance, family *Mymonaviridae* belongs to the order *Mononegavirales*, as are viruses that cause Ebola, measles, mumps, rabies and respiratory diseases. Families *Metaviridae* and *Pseudoviridae* belong to order *Ortervirales*, together with human immunodeficiency virus (HIV), cause of acquired immunodeficiency syndrome (AIDS), and other retroviruses.

Classification of exemplar mycoviruses known to infect *Aspergillus* spp is shown in **Figure 2**.

Almost all known mycoviruses have double stranded (ds) RNA genomes or single stranded (ss) RNA genomes, either positive sense or negative sense, with one family of mycoviruses having circular ssDNA genomes. Virions are often proteinaceous in nature, composed of virus capsid proteins and their structure may range from spherical, to bacilliform in the case of barnaviruses, to filamentous in the case of flexiviruses and mymonaviruses. The absence of true virions is also common: narnaviruses and mitoviruses exist as naked RNA molecules respectively in the cytoplasm and mitochondria, hypoviruses are encapsulated in host derived lipid vesicles, polymycoviruses are non-conventionally encapsidated by a viral protein [66, 67]. Mycoviruses move intracellularly within the infected fungus and spread in mycelia during cell division and growth. Almost all known mycoviruses lack an extracellular phase in their replication cycle; they are transmitted vertically during asexual and/or sexual spore production and horizontally between fungal strains following cell fusion. The absence of an extracellular phase explains the general lack of lipid envelopes in virions.

Early reports focused on the mycovirus-mediated alterations on fungal phenotype, including morphology, pigmentation, asexual and sexual sporulation, and growth. Production of viral toxins conferring a competitive advantage to the fungal host [68], clearly illustrate that viral infection can be beneficial to the host and viruses are undeserving of their name, derived from the Latin word for 'poison' or 'venom'. These killer yeast systems have been primarily studied in the eukaryotic model organism *Saccharomyces cerevisiae* [69], extensively used in biotechnological applications such as baking, brewing and winemaking. However, interest in mycoviruses stems mainly from their effects on the interaction between their host fungus and the plant, insect or mammalian/human host of the fungus.

An increasing number of studies clearly illustrate the importance of mycoviruses in host-microbe interactions. The discovery of 'transmissible hypovirulence', i.e., mycovirus-mediated decrease in fungal pathogenicity represents a major advance in the field and the first mycovirus-based biological control application to combat chestnut blight caused by the plant pathogen *Cryphonectria parasitica* [70, 71]. The opposite phenomenon called hypervirulence, i.e., mycovirus-mediated increase in fungal pathogenicity, has also been noted. For instance, two variants of Aspergillus fumigatus polymycovirus 1 (AfuPmV-1), the first virus demonstrated to be infectious as dsRNA [66], respectively cause hypovirulence in an immunosuppressed mouse infection model [72] and hypervirulence in the greater wax moth *G. mellonella* infection model [73]. Additionally, AfuPmV-1 renders its fungal host more

REALM	KINGDOM	CLASS	ORDER	FAMILY	GENUS

Riboviria

→ **Polymycoviridae** – DS

→ *Orthornavirae*

→ **Botybirnavirus** – DS ⬡

→ *Duplornaviricota*

→ *Chrymotiviricetes*

↳ *Ghabrivirales*

→ **Chrysoviridae** – DS ⬡
→ **Megabirnaviridae** – DS ⬡
→ **Quadriviridae** – DS ⬡
→ **Totiviridae** – DS ⬡

↳ *Resentoviricetes*

↳ *Reovirales*

→ *Kitrinoviricota*

↳ *Alsuviricetes*

→ **Reoviridae** – DS ⬡

↳ *Tymovirales*

→ **Alphaflexiviridae** – (+)SS ▬▬▬▬
→ **Deltaflexiviridae** – (+)SS ▬▬▬
→ **Gammaflexiviridae** – (+)SS ▬▬▬

→ *Martellivirales*

→ *Lenarviricota*

↳ **Endornaviridae** – (+)SS

↳ *Amabiliviricetes*

↳ *Wolframvirales*

→ **Narnaviridae** – (+)SS

→ *Howeltoviricetes*

↳ *Cryppavirales*

↳ **Mitoviridae** – (+)SS

→ *Miaviricetes*

↳ *Ourlivirales*

→ *Negarnaviricota*

↳ **Botourmiaviridae** – (+)SS

↳ *Monjiviricetes*

↳ *Mononegavirales*

→ *Pisuviricota*

↳ **Mymonaviridae** – (–)SS ▬▬▬▬

→ *Duplopiviricetes*

↳ *Durnavirales*

→ **Amalgaviridae** – DS
→ **Hypoviridae** – (+)SS
→ **Partitiviridae** – DS ⬡

→ *Pisoniviricetes*

↳ *Sobelivirales*

→ *Pararnavirae*

↳ *Artverviricota*

→ **Barnaviridae** – (+)SS ⬡

→ *Revtraviricetes*

↳ *Ortervirales*

→ **Metaviridae** – (+)SS ●
→ **Pseudoviridae** – (+)SS ●

Monodnaviria

↳ *Shotokuvirae*

↳ *Cressdnaviricota*

↳ *Repensiviricetes*

↳ *Geplafuvirales*

↳ **Genomoviridae** – (+)SS ⬡

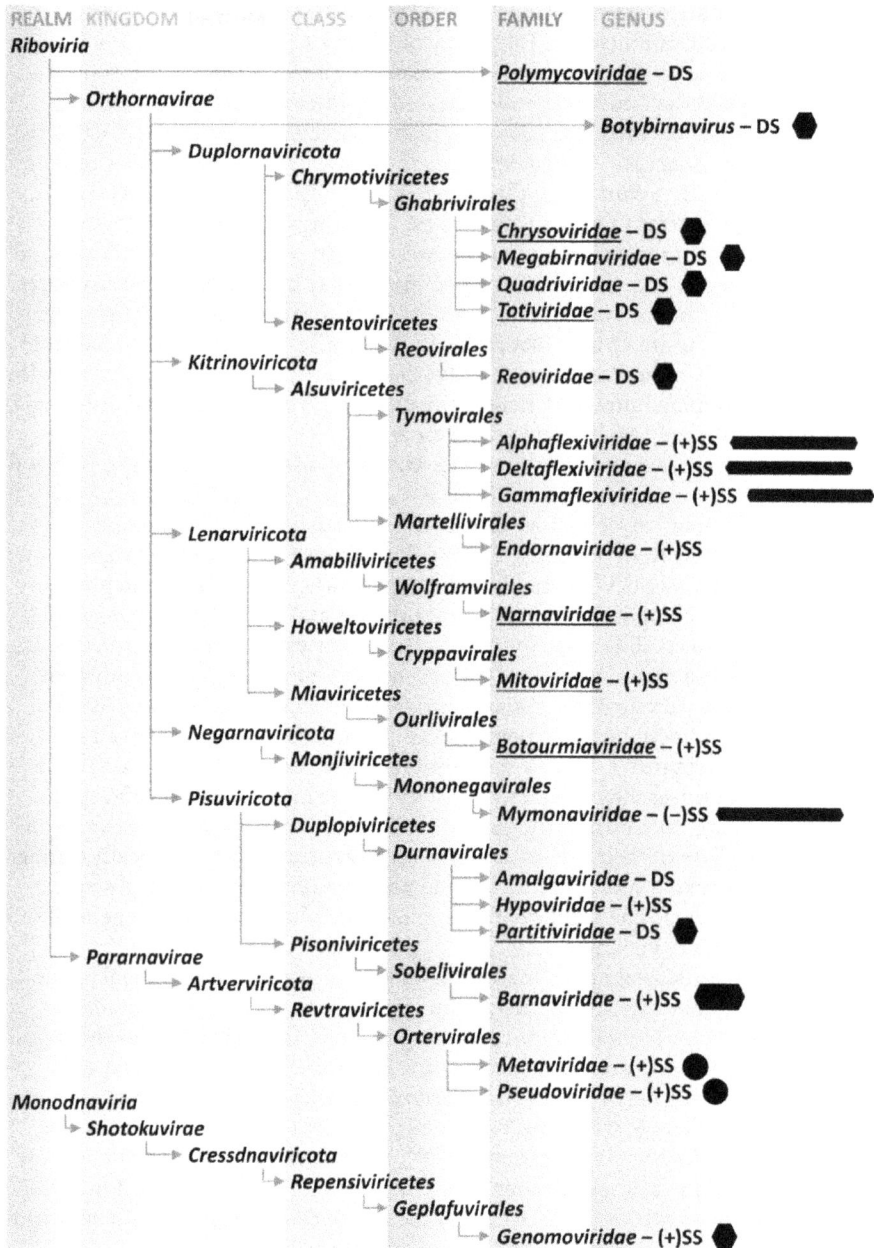

Figure 1.
Current classification of mycoviruses according to the International Committee on Taxonomy of Viruses. The realms Riboviria *and* Monodnaviria *accommodate viruses with respectively RNA and DNA genomes. Underlying family names accommodate mycoviruses known to infect* Aspergillus spp. *Next to family/genus names, (+)SS, (–)SS and DS indicate respectively, positive-sense single-stranded, negative-sense single-stranded and double-stranded genomes; hexagons indicate the presence of true virions, either isometric, bacilliform of filamentous.*

sensitive to the bacterium *Pseudomonas aeruginosa* [74]. Furthermore, partitivirus infection of *Talaromyces marneffei* leads to hypervirulence in a BALB/c mouse model [75]. Mycoviruses dsRNA genomes or replication intermediates are recognized by Toll-like receptor 3 (TLR-3) [76] and may induce an interferon immune response

FAMILY	GENUS	VIRUSES
Polymycoviridae		
	Polymycovirus	
		Aspergillus fumigatus polymycovirus 1
		Aspergillus fumigatus polymycovirus 1M
		Aspergillus spelaeus polymycovirus 1
Chrysoviridae		
	Alphachrysovirus	
		Aspergillus fumigatus chrysovirus
	Betachrysovirus	
		Aspergillus thermomutatus chrysovirus 1
Totiviridae		
	Victorivirus	
		Aspergillus foetidus slow virus 1
Narnaviridae		
	Narnavirus	
		Aspergillus fumigatus narnavirus 1
		Aspergillus fumigatus narnavirus 2
Mitoviridae		
	Mitovirus	
		Aspergillus fumigatus mitovirus 1
Boturmiaviridae		
	Penoulivirus	
		Aspergillus neoniger ourmia-like virus 1
Partitiviridae		
	Gammapartitivirus	
		Aspergillus ochraceous partitivirus
		Aspergillus fumigatus partitivirus 1
		Aspergillus fumigatus partitivirus 2

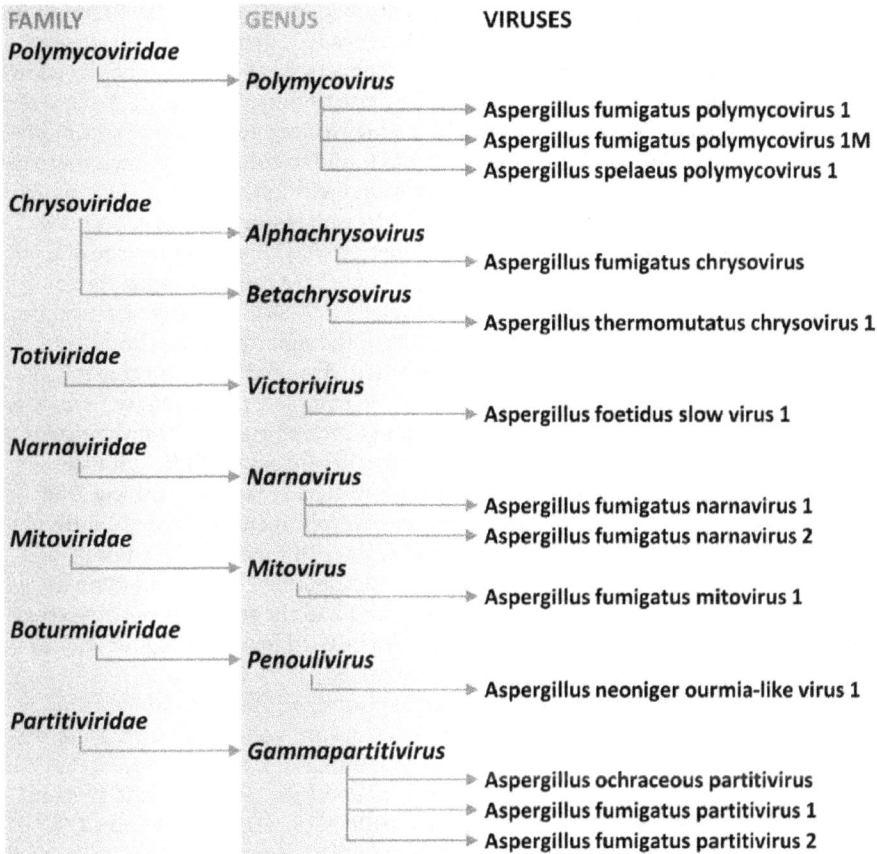

Figure 2.
Classification of exemplar mycoviruses known to infect Aspergillus *spp. Not all known mycoviruses found in* Aspergillus *spp. are officially assigned to recognized taxa. The phenotypes and effects of the majority of these mycoviruses on their* Aspergillus *host is unknown.*

in a TLR-3 dependent or independent manner, as illustrated with totivirus infected *Malassezia* [77, 78]. A link between azole resistance and mycovirus infection has been noted in *Penicillium digitatum* [79]. Finally, mycovirus infection is known to be responsible for modulation of fungal toxins and this phenomenon has been studied mainly in *Aspergillus* spp [80]. Carcinogenic aflatoxin production may be repressed by the presence of a mycovirus in *A. flavus* [81–84], while ochratoxin A is enhanced by the presence of a partitivirus in *A. ochraceus* [85].

Currently most mycovirus studies are focused on economically important phytopathogenic fungi, while scant data regarding fungi containing mycoviruses and human disorders are available. Since mycoviruses do exist in fungi, and humans are exposed to them, further research on these organisms may expand our knowledge of their possible role and effects of their interaction with humans.

5. Studies of mycovirus containing *Aspergillus flavus*

A report describing plasma of patients with acute lymphoblastic leukemia (ALL) having a positive reaction to an *A. flavus* isolate containing an unknown mycovirus is available [86]. Exposure of the peripheral blood mononuclear cells

(PBMCs) obtained from a group of ALL patients who were in a complete remission to the culture of this organism was reported to reproduce genetic and cell surface phenotypes, characteristic of active ALL [87]. Conversely, this was not observed in the control group of patients [87]. To describe these findings (which are patented) in more detail, in a series of experiments, a mycovirus infected *A. flavus* separated from the home of a patient with B-cell ALL was found to contain unknown mycovirus particles. These mycovirus particles were found within the body of the organism and culture supernatant. Chemical analysis of the isolated mycovirus containing *A. flavus* had revealed a lack of aflatoxin production [86]. The latter may be due to the influence of the unknown mycovirus which may have caused suppression of the production of aflatoxin as described previously [80–84]. Utilizing fast protein liquid chromatography (FPLC) for the analysis of the supernatant of the culture of this isolate, three separate peaks were identified. As noted above, in controlled experiments using plasma of patients with ALL in complete remission, with no evidence of the disease, using crude supernatant of the culture of the mycovirus containing *A. flavus* and enzyme-linked immunosorbent assay (ELISA) for the detection of antibodies, plasma of patients with ALL had reacted positively. The plasma obtained from three separate groups of controls, including normal individuals, patients with sickle cell disease and individuals with various solid tumors, had been negative. In a separate study evaluating peaks obtained by fractionation using FPLC, of the three peaks which were found, peak 1 had the strongest positive effect [86]. The authors suggest that this technique can be used for screening for ALL or a test to identify patients who have had this disease [86].

As noted before, in a related publication, exposure of PBMCs obtained from ALL patients in complete remission, and long-term survivors of this disease, to the supernatant of the culture of the mycovirus containing *A. flavus* resulted in the re-development of the genetic and cell surface phenotypes, characteristic of ALL. The cell surface phenotypes examined were CD10/CD19, CD19/CD34 and CD34/CD117. The redevelopment of the ALL cell surface phenotypes was reported to be gradual, completed in 24 hours, and remained stable thereafter. Following exposure to the supernatant of the mycovirus containing *A. flavus*, alterations in gene expression were evaluated using microarray technique. Some of these alterations were reported to be upregulation of JAK1 (12.87-fold), JAK2 (1.5-fold), JAK3 (2.73-fold), IKZF1 (10.12-fold), MCL1 (59.37-fold), MYC (14.19-fold), HDAC1 (26.39-fold) and downregulation of PAX5 (3.05-fold). Following incubation, a significant and robust activation of transcription factor NF-$\varkappa$B p65 was reported by immunoblotting in ALL patients without any changes in the controls. The supernatant of the culture of *Mycocladus corymbifer*, which was used as a negative control, was reported to have no effects on PBMCs either from the ALL or control patients [87]. The above studies suggest a possible role for the mycovirus containing *A. flavus* in the process of leukemogenesis and opens a venue for vaccination and prevention of this disease.

6. Conclusion

It is apparent that fungal *spp.* are important in human and animal health. The mechanism of the effects of fungal agents in the development of human diseases appears to be multifaceted. Fungi are widespread in nature and inevitably, humans encounter these organisms. Many fungi contain mycoviruses. Although a significant amount of data regarding the carcinogenic effects of mycotoxins in the development of malignant disorders are available, possible pathogenicity and role of the mycoviruses in fungi, if any, in human and animal health, including malignant disorders, are not known. Recent reports describing *in vitro* effects of a mycovirus containing *A. flavus* isolate in redeveloping characteristic ALL cell surface and genetic phenotypes in the PMBCs of acute lymphoblastic leukemia patients in complete remission is of interest. The existence of antibody to this organism in plasma of these patients is intriguing and further indicates its possible role in leukemogenesis. This area needs to be further investigated.

Author details

Cameron K. Tebbi[1*], Ioly Kotta-Loizou[2] and Robert H.A. Coutts[3]

1 Children's Cancer Research Group Laboratory, Tampa, Florida, USA

2 Department of Life Sciences, Imperial College London, London, United Kingdom

3 School of Medical and Life Sciences, University of Hertfordshire, Hatfield, United Kingdom

*Address all correspondence to:
ctebbi@childrenscancerresearchgrouplaboratory.org

IntechOpen

References

[1] Jovanovic, S, Felder-Kennel, A, Gabrio, T, Kouros, B, Link, B, Maisner, V, Piechotowski, I, Schick, K-H, Schrimpf, M, Ursula Weidner, M, Zöllner, I, Schwenk, M. Indoor fungi levels in homes of children with and without allergy history. *Int J Hyg Environ Health* **207**:369-378, 2004.

[2] Van Burik, J, Magee, PT. Aspects of fungal pathogenesis in humans. *Annu Rev Microbiol* **55**:743-772, 2001.

[3] McGinnis, MR, Sigler L, Rinaldi MG. Some medically important fungi and their common synonyms and names of uncertain application. *Clin Infect Dis* **29**:728-730, 1999.

[4] Sternberg, S. The emerging fungal threat. *Science* **266**:1632-1634,1994.

[5] Barrios, N, Tebbi, CK, Rotstein, C, Siddiqui, S, Humbert, JR. Brainstem invasion by *Aspergillus fumigatus* in a child with leukemia. *NY State J Med* **88**:656-658, 1988.

[6] Rotstein, C, Tebbi, CK, Brass, C. Viral, bacterial and fungal infections in adolescent oncology. In *Adolescent Oncology*, Tebbi CK (Editor), Futura Publishing Company, Mt. Kisco, NY, 1987, 429-506.

[7] Brown, GD, Denning, DW, Gow, NA, Levitz, SM, Netea, MG, White, TC. Hidden killers: human fungal infections. *Sci Transl Med* **4**:165rv13, 2012.

[8] Types of Fungal Diseases. Centers for Disease Control and Prevention, National Center for Emerging and Zoonotic Infectious Diseases (NCEZID), Division of Foodborne, Waterborne, and Environmental Diseases (DFWED), USA, May 6, 2019.

[9] Aykut, B, Pushalkar, S, Chen, R, Li Q, Abengozar, R, Kim, JI, Shadaloey, SA, Wu, D, Preiss, P, Verma, N, Guo, Y, Saxena, A, Vardhan, M, Diskin, B, Wang, W, Leinwand, J, Kurz, E, Kochen Rossi, JA, Hundeyin, M, Zambrinis, C, Li, X, Saxena, D, Miller, G. The fungal mycobiome promotes pancreatic oncogenesis via activation of MBL. *Nature* **574**:264-267, 2019.

[10] Ramirez-Garcia, A, Rementeria, A, Aguirre-Urizar, JM, Moragues, MD, Antoran A, Pellon, A, Abad-Diaz-de-Cerio, A, Hernando, FL. *Candida albicans* and cancer: Can this yeast induce cancer development or progression? *Crit Rev Microbiol* **42**:181-93, 2016.

[11] Sankari, SL, Gayathri, K, Balachander, N, Malathi, L. *Candida* in potentially malignant oral disorders. *J Pharm Bioallied Sci* **7**:S162-164, 2015.

[12] Nørgaard, M, Thomsen, RW, Farkas, DK, Mogensen, MF, Sørensen, HT. Candida infection and cancer risk: a Danish nationwide cohort study. *Eur J Intern Med* **24**:451-455, 2013.

[13] Bennett, JW, Klich, M. Mycotoxins. *Clin Microbiol Rev* **16**:497-516, 2003.

[14] Bennett, JW. Mycotoxins, mycotoxicoses, mycotoxicology and mycopathologia. *Mycopathologia* **100**:3-5, 1987.

[15] Iqbal, SZ, Nisar, S, Asi, MR, Jinap, S. Natural incidence of aflatoxins, ochratoxin A and zearalenone in chicken meat and eggs. *Food Control*, **43**:98-103, 2014.

[16] Khlangwiset P, Shephard GS, Wu F. Aflatoxins and growth impairment: a review. *Crit Rev Toxicol* **41**:740-55, 2011.

[17] Boonen J, Malysheva SV, Taevernier L, Diana Di Mavungu J, De Saeger S, De Spiegeleer B. Human skin

penetration of selected model mycotoxins. *Toxicology* **301**:21-32, 2012.

[18] Njoroge, SMC, Matumba, L, Kanenga, K, Siambi, M, Waliyar, F, Maruwo, J, Machinjiri, N, Monyo, ES. Aflatoxin B1 levels in groundnut products from local markets in Zambia. *Mycotoxin Res* 33:113-119, 2017.

[19] Lioi, M, Santoro, A, Barbieri, R, Salzano,S, Ursini, M. Ochratoxin A and zearalenone: a comparative study on genotoxic effects and cell death induced in bovine lymphocytes. *Mutat Res* **557**:19-27, 2004.

[20] Muller, G, Burkert, B, Moller, U, et al. Ochratoxin A and some of its derivatives modulate radical formation of porcine blood monocytes and granulocytes. *Toxicology* **199**:251-259, 2004.

[21] Muller, G, Rosner, H, Rohrmann, B et al. Effects of the mycotoxin ochratoxin A and some of its metabolites on the human cell line THP-1. *Toxicology* **184**: 69-82, 2003.

[22] Nielsen, KF, Smedsgaard, J. Fungal metabolite screening: database of 474 mycotoxins and fungal metabolites for dereplication by standardised liquid chromatography–UV–mass spectrometry methodology. *J Chromatogr A* **1002**:111-136, 2003.

[23] Müller, A, Lehmann, I, Seiffart, A, Diez, U, Wetzig, H, Borte, M, Herbarth, O. Increased incidence of allergic sensitization and respiratory diseases due to mold exposure: Results of the Leipzig Allergy Risk children Study (LARS). *Int J Hyg Environ Health* **204**:363-365, 2002.

[24] Romagnani S. Lymphokine production by human T cells in disease states. *Annu Rev Immunol* **12**:227-257, 2003.

[25] Nutt, SL, Huntington, ND. Cytotoxic T lymphocytes and natural killer cells. In *Clinical Immunology: Principles and Practice*, Rich RR, Fleisher TA, Shearer WT, Schroeder HW, Frew AJ, Weyand CM (Editors), Fifth Edition, Elsevier Publications 2019, 247-259.

[26] Barrett JR. Mycotoxins: of molds and maladies. *Environ Health Perspect* **108**:A20-A23, 2000.

[27] Dhakal, A, Sbar, E. Aflatoxin toxicity. In StatPearls [Internet]. Treasure Island (FL): StatPearls Publishing; 2021.

[28] *Nogueira, L, Foerster, C, Groopman, J, Egner, P, Koshiol, J, Ferreccio, C.* Association of aflatoxin with gallbladder cancer in Chile. *JAMA* **313**:*2075-2077, 2015.*

[29] Barrett, JR. Liver cancer and aflatoxin: New information from the Kenyan outbreak. *Environ Health Perspect* **113**:A837-A838, 2005.

[30] Benkerroum, N. Chronic and acute toxicities of aflatoxins: Mechanisms of action. *Int. J. Environ. Res. Public Health* **17**:423, 2020.

[31] *Chen, C, Mitchell, NJ, Gratz, J, Houpt, ER, Gong, Y, Egner, PA, Groopman, JD, Riley, RT, Showker, JL, Svensen, E, Mduma, ER, Patil, CL, Wu, F.* Exposure to aflatoxin and fumonisin in children at risk for growth impairment in rural Tanzania. *Environ Int* **115**:*29-37, 2018.*

[32] *Mitchell NJ, Hsu HH, Chandyo RK, Shrestha B, Bodhidatta L, Tu YK, Gong YY, Egner PA, Ulak M, Groopman JD, Wu F.* Aflatoxin exposure during the first 36 months of life was not associated with impaired growth in Nepalese children: An extension of the MAL-ED study. *PLOS ONE* **12**:*e0172124, 2017.*

[33] Turner PC, Collinson AC, Cheung YB, Gong Y, Hall AJ, Prentice AM, Wild CP. Aflatoxin exposure in utero causes growth faltering in Gambian infants. *Int J Epidemiol* **36:**1119-1125, 2007.

[34] Williams, JH, Phillips, TD, Jolly, PE, Stiles, JK, Jolly, CM, Aggarwal, D. Human aflatoxicosis in developing countries: a review of toxicology, exposure, potential health consequences, and interventions. *Am J Clin Nutr* **80:**1106-1122, 2004.

[35] Squire, RA. Ranking animal carcinogens: a proposed regulatory approach. *Science* **214:**877-880, 1981.

[36] Deng ZL, Ma Y. Aflatoxin sufferer and p53 gene mutation in hepatocellular carcinoma. *World J Gastroenterol* **4:**28-29, 1998.

[37] Aguilar F, Hussain SP, Cerutti P. Aflatoxin B1 induces the transversion of G-->T in codon 249 of the p53 tumor suppressor gene in human hepatocytes. Proc Natl Acad Sci *USA* **90:**8586-8590, 1993.

[38] Umeda M, Tsutsui T, Saito M. Mutagenicity and inducibility of DNA single-strand breaks and chromosome aberrations by various mycotoxins. *Gan* **68:**619-625, 1977.

[39] Wray, BB, Harmon, CA, Rushing, EJ, Cole, RJ. Precipitins to an aflatoxin-producing strain of *Aspergillus flavus* in patients with malignancy. *J Cancer Res Clin Oncol* **103:**181-185, 1982.

[40] Wray, BB, Rushing, EJ, Boyd, RC, Schindel, AM Suppression of phytohemagglutinin response by fungi from a "leukemia" house. *Arch Environ Health* **34:**350-353, 1979.

[41] Wray, BB, O'Steen, KG Mycotoxin-producing fungi from house associated with leukemia. *Arch Environ Health* **30:**571-573, 1975.

[42] McPhedran, P, Heath, CW. Multiple cases of leukemia associated with one house. *JAMA* **209:**2021-2025, 1969.

[43] Mesri EA, Feitelson MA, Munger K. Human viral oncogenesis: a cancer hallmarks analysis. *Cell Host Microbe* **15:**266-282, 2014.

[44] Momin B, Richardson L. An analysis of content in comprehensive cancer control clans that address chronic hepatitis B and C virus infections as major risk factors for liver cancer. *J Community Health* **37:**912-916, 2012.

[45] Snow AN, Laudadio J. Human papilloma virus detection in head and neck squamous cell carcinomas. *Adv Anat Pathol* **17:**394-403, 2010.

[46] Liao JB. Viruses and human cancer. *Yale J Biol Med* **79:**115-122, 2006.

[47] Montaner, S, Sodhi, A, Ramsdell, AK, Martin, D, Hu, J, Sawai, ET, Gutkind, JS. The Kaposi's sarcoma-associated herpesvirus G protein-coupled receptor as a therapeutic target for the treatment of Kaposi's sarcoma. *Cancer Res* **66:**168-174, 2006.

[48] Lehtinen M, Koskela P, Ogmundsdottir HM, Bloigu A, Dillner J, Gudnadottir M, Hakulinen T, Kjartansdottir A, Kvarnung M, Pukkala E, Tulinius H, Lehtinen T. Maternal herpesvirus infections and risk of acute lymphoblastic leukemia in the offspring. *Am J Epidemiol* 158:207-213, 2003.

[49] Sarid R, Olsen SJ, Moore PS. Kaposi's sarcoma-associated herpesvirus: epidemiology, virology, and molecular biology. *Adv Virus Res* **52:**139-232, 1999.

[50] Flore, O, Rafii, S, Ely, S, O'Leary, JJ, Hyjek, EM, Cesarman, E. Transformation of primary human endothelial cells by Kaposi's

sarcoma-associated herpesvirus. *Nature* **394**:588-592, 1998.

[51] Chang, Y, Cesarman, E, Pessin, MS, Lee, F, Culpepper, J, Knowles, DM, Moore, PS. Identification of herpesvirus-like DNA sequences in AIDS-associated Kaposi's sarcoma. *Science* **266**:1865-1869, 1994.

[52] Gold, JE, Castella,A, Zalusky, R. B-cell acute lymphoblastic leukemia in HIV antibody-positive patients. *J Hematol* **32**:200-204, 1989.

[53] Tebbi, CK. Etiology of acute leukemia: A review. *Cancers* **13**:2256, 2021.

[54] Rowe, M, Fitzsimmons, L, Bell, AI. Epstein-Barr virus and Burkitt lymphoma. *Chin J Cancer* **33**:609-619, 2014.

[55] Haque T, Wilkie GM, Taylor C, et al. Treatment of Epstein-Barr-virus-positive post-transplantation lymphoproliferative disease with partly HLA-matched allogeneic cytotoxic T cells. *Lancet* **360**:436-442, 2002.

[56] Gottschalk S, Gottschalk S, Ng CY, Perez M, Smith CA, Sample C, Brenner MK, Heslop HE, Rooney CM. An Epstein-Barr virus deletion mutant associated with fatal lympho proliferative disease unresponsive to therapy with virus-specific CTLs. *Blood* **97**:835-843, 2001.

[57] Papadopoulos EB, Ladanyi, M, Emanuel, D et al. Infusions of donor leukocytes to treat Epstein-Barr virus-associated lymphoproliferative disorders after allogeneic bone marrow transplantation. N Engl J Med **330**:1185-1191, 1994.

[58] Thorley-Lawson DA, Poodry CA. Identification and isolation of the main component (gp350-gp220) of Epstein-Barr virus responsible for generating neutralizing antibodies in vivo. *J Virol* **43**:730-736, 1982.

[59] Ho JH. An epidemiologic and clinical study of nasopharyngeal carcinoma. *Int J Radiat Oncol Biol Phys* **4**:182-198, 1978.

[60] Kaufmann, AM, Stern, PL, Rankin, EM, Sommer, H, Nuessler, V, Schneider, A, Adams, M, Onon, TS, Bauknecht, T, Wagner, U, Kroon, K, Hickling, J, Boswell, CM, Stacey SN, Kitchener, HC, Gillard, J, Wanders, J, Roberts, JS, Zwierzina, H. Safety and immunogenicity of TA-HPV, a recombinant vaccinia virus expressing modified human papillomavirus (HPV)-16 and HPV-18 E6 and E7 genes, in women with progressive cervical cancer. *Clin Cancer Res* **8**:3676-3685, 2002.

[61] Wallin, KL, Wiklund, F, Angström, T, Bergman, F, Stendahl, U, Wadell, G, Hallmans, G, Dillner, J.Type-specific persistence of human papillomavirus DNA before the development of invasive cervical cancer. *N Engl J Med* **341**:1633-1638, 1999.

[62] von Knebel Doeberitz, M, Oltersdorf, T, Schwarz, E, Gissmann, L. Correlation of modified human papilloma virus early gene expression with altered growth properties in C4-1 cervical carcinoma cells. *Cancer Res* **48**:3780-3786, 1988.

[63] Halpert, R, Fruchter, RG, Sedlis, A, Butt, K, Boyce, JG, Sillman, FH. Human papillomavirus and lower genital neoplasia in renal transplant patients. *Obstet Gynecol* **68**:251-258, 1986.

[64] Kotta-Loizou, I. Mycoviruses: past, present, and future. *Viruses* **11**:361, 2019.

[65] Krupovic, M, Ghabrial, SA, Jiang, D, Varsani, A. *Genomoviridae*: a new family of widespread single-stranded DNA viruses. *Arch Virol* **161**:2633-2643, 2016.

[66] Kanhayuwa, L, Kotta-Loizou, I, Özkan, S, Gunning, AP, Coutts, RHA. A novel mycovirus from *Aspergillus fumigatus* contains four unique dsRNAs as its genome and is infectious as dsRNA. *Proc Natl Acad Sci U S A* **112**:9100-9105, 2015.

[67] Kotta-Loizou I, Coutts, RHA. Studies on the virome of the entomopathogenic fungus *Beauveria bassiana* reveal novel dsRNA elements and mild hypervirulence. *PLoS Pathog* **13**:e1006183, 2017.

[68] Drinnenberg, IA, Fink, GR, Bartel, DP. Compatibility with killer explains the rise of RNAi-deficient fungi. *Science* **333**:1592, 2011.

[69] Schmitt, MJ, Breinig, F. Yeast viral killer toxins: lethality and self-protection. *Nat Rev Microbiol* **4**:212-221, 2006.

[70] Van Alfen, NK, Jaynes, RA, Anagnostakis, SL, Day, PR. Chestnut blight: biological control by transmissible hypovirulence in *Endothia parasitica*. *Science* **189**:890-891, 1975.

[71] Anagnostakis, SL. Biological control of chestnut blight. *Science* **215**:466-471, 1982.

[72] Takahashi-Nakaguchi, A, Shishido, E, Yahara, M, Urayama, SI, Ninomiya, A, Chiba, Y, Sakai, K, Hagiwara, D, Chibana, H, Moriyama, H, Gonoi, T. Phenotypic and molecular biological analysis of polymycovirus AfuPmV-1M from *Aspergillus fumigatus*: reduced fungal virulence in a mouse infection model. *Front Microbiol* **11**:607795, 2020.

[73] Özkan, S, Coutts, RHA. *Aspergillus fumigatus* mycovirus causes mild hypervirulent effect on pathogenicity when tested on *Galleria mellonella*. *Fungal Genet Biol* **76**:20-26, 2015.

[74] Nazik, H, Kotta-Loizou, I, Sass, G, Coutts, RHA, Stevens, DA. Virus infection of *Aspergillus fumigatus* compromises the fungus in intermicrobial competition. *Viruses* **13**:686, 2021.

[75] Lau, SKP, Lo, GCS, Chow, FWN, Fan, RYY, Cai, JJ, Yuen, KY, Woo, PCY. Novel partitivirus enhances virulence of and causes aberrant gene expression in *Talaromyces marneffei*. *mBio* **9**:e00947-18, 2018.

[76] Ives, A, Ronet, C, Prevel, F, Ruzzante, G, Fuertes-Marraco, S, Schutz, F, Zangger, H, Revaz-Breton, M, Lye, LF, Hickerson, SM, Beverley, SM, Acha-Orbea, H, Launois, P, Fasel, N, Masina, S. Leishmania RNA virus controls the severity of mucocutaneous leishmaniasis. *Science* **331**:775-778, 2011.

[77] Park, M, Cho, YJ, Kim, D, Yang, CS, Lee, SM, Dawson, TL Jr, Nakamizo, S, Kabashima, K, Lee, YW, Jung, WH. A novel virus alters gene expression and vacuolar morphology in *Malassezia* cells and induces a TLR3-mediated inflammatory immune response. *mBio* **11**:e01521-20, 2020.

[78] Applen Clancey, S, Ruchti, F, LeibundGut-Landmann, S, Heitman, J, Ianiri, G. A novel mycovirus evokes transcriptional rewiring in the fungus *Malassezia* and stimulates beta interferon production in macrophages. *mBio* **11**:e01534-20, 2020.

[79] Niu, Y, Yuan, Y, Mao, J, Yang, Z, Cao, Q, Zhang, T, Wang, S, Liu, D. Characterization of two novel mycoviruses from *Penicillium digitatum* and the related fungicide resistance analysis. *Sci Rep* **8**:5513, 2018.

[80] Kotta-Loizou, I, Coutts, RHA. Mycoviruses in *Aspergilli*: A comprehensive review. *Front Microbiol* **8**:1699-1714, 2017.

[81] Schmidt, FR. The RNA interference-virus interplay: tools of nature for gene modulation, morphogenesis, evolution

and a possible mean for aflatoxin control. *Appl Microbiol Biotechnol* **83:**611-615, 2009.

[82] Silva, VN, Durigon, EL, de Fátima Costa Pires, M, Lourenço, A, de Faria, MJ, Corrêa, B. Time course of virus-like particles (VLPs) double-stranded RNA accumulation in toxigenic and non-toxigenic strains of *Aspergillus flavus*. *Braz J Microbiol* **32:**56-60, 2001.

[83] Schmidt, FR, Lemke, PA, and Esser, K. Viral influences on aflatoxin formation by *Aspergillus flavus*. *Appl Microbiol Biotechnol* **24:**248-252, 1986.

[84] Schmidt, FR, Davis, ND, Diener, UL and Lemke, PA. Cycloheximide induction of aflatoxin synthesis in a nontoxigenic strain of *Aspergillus flavus*. BioTechnology **1:**794-795, 1983.

[85] Nerva, L, Chitarra, W, Siciliano, I, Gaiotti, F, Ciuffo, M, Forgia, M, Varese, GC, Turina, M. Mycoviruses mediate mycotoxin regulation in *Aspergillus ochraceus*. *Environ Microbiol* **21:**1957-1968, 2019.

[86] Tebbi, CK, Badiga, A, Sahakian, E, Arora, AI, Nair, S, Powers, JJ, Achille, AN, Jaglal, MV, Patel, S, Migone, F. Plasma of acute lymphoblastic leukemia patients react to the culture of a mycovirus containing *Aspergillus flavus*. *J Pediatr Hematol Oncol* **42:**350-358, 2020.

[87] Tebbi, CK, Badiga, A, Sahakian, E, Powers, JJ, Achille, AN, Patel, S, Migone, F. Exposure to a mycovirus containing *Aspergillus flavus* reproduces acute lymphoblastic leukemia cell surface and genetic markers in cells from patients in remission and not controls. *Cancer Treat Res Commun* **26:**100279, 2020.

Chapter 6

Industrial Applications of Nanomaterials Produced from *Aspergillus* Species

Mahendra Rai, Indarchand Gupta, Shital Bonde,
Pramod Ingle, Sudhir Shende, Swapnil Gaikwad,
Mehdi Razzaghi-Abyaneh and Aniket Gade

Abstract

There is a great demand for green methods of synthesis of nanoparticles. Fungi play an important role in the synthesis of nanoparticles, of which *Aspergillus* spp. are known to secrete different enzymes responsible for the synthesis of nanoparticles. The process of biosynthesis of nanoparticles is simple, rapid, cost-effective, eco-friendly, and easy to synthesize at ambient temperature and pressure. Mostly, the metal nanoparticles such as silver, gold, lead and the oxides of titanium, zinc, and copper are synthesized from *Aspergillus* spp. These include mainly *Aspergillus fumigatus, A. flavus, A. niger, A. terreus*, and *A. clavatus*. The fabrication of different nanoparticles is extracellular. In the present chapter, we have discussed the role of different species of *Aspergillus*, mechanism of biogenic synthesis particularly enzymes involved in the reduction of metal ions into nanoparticles. The biogenically synthesized nanoparticles have demonstrated several biomedicals, agricultural, and engineering applications. The biogenic nanoparticles are mostly used as antimicrobial and cytotoxic agents. Their use as fungicidal agents is important for sustainable agriculture.

Keywords: *aspergillus* spp., nanomaterials, biogenic synthesis, industrial, biomedical, agriculture

1. Introduction

Nanomaterials (NMs) are the structures fabricated in the nanoscale, i.e. 1 to 100 nm and having at least one dimension in the nanoscale. The fabrication, study, and application of nanostructures are known as nanotechnology. The exhibition of novel physicochemical properties by the nanoscale materials has provided a unique opportunity for researchers to design and develop materials with applications in the diverse fields of science and technology. This has attracted attention towards nanoparticles (NPs) and their fabrication as compared to other sectors of NMs. Some of the nanomaterial productions have reached to the industrial scale due to the high demand for NMs in consumer products and their number is increasing at the moment with their developing applications. Ever-increasing demand for different NPs has generated the need for easy, safe, efficient, rapid, and eco-friendly procedures for their large-scale production.

Nanomaterials can be produced by two general approaches, i.e. top-down approach and bottom-up approach. Another classification includes different methods like physical, chemical, biological, and hybrid methods of nanoparticle production. The physical method requires an expensive setup, is high energy-consuming, and hazardous to health and the environment. Whereas chemical methods are highly efficient as compared to physical methods, but involve a toxic reducing agent, solvent, and stabilizing/capping agents. Recently, the biological method of nanoparticle production has attracted attention because of its ease, eco-friendly nature, high efficiency, and high yield. In this method, a biological agent or a biomolecule plays a significant role in the production of NMs [1]. Production of NMs by a biological method is a promising alternative for physical and chemical methods [2].

Among the different biological systems like bacteria, actinomycetes, fungi, plants, protozoa, and animals, fungi have shown great potential for the production of NPs on large scale. Bacteria normally produced NPs intracellularly, where large-scale production and purification of NPs is complicated and expensive. Unlike bacteria, fungi produce NPs extracellularly and are easy to use and purify NPs for large-scale production [3]. Fungi are easy to handle, versatile, tolerant, and economical biological systems for industrial production of biotechnology products and have been used extensively in large-scale production of different metabolites. The tremendous ability of fungi in the secretion of proteins up to 100 g/L, metabolic diversity, and high production capacity have made them a unique option for industrial biotechnology for decades. Hence, filamentous fungi are the first choice, since they are capable of secreting a large amount of proteins and other metabolites extracellularly. Moreover, the fabrication of NPs by a fungal system is a green process [4]. Among the fungal sources, *Aspergillus* is a very promising candidate for the production of NPs, this is because there are more than 350 species of this genus with enormous biochemical versatility in addition to the secretion of a large quantity of proteins [5]. Different *Aspergillus* species produce NPs of diverse sizes and shapes with interesting physicochemical properties like enhanced thermostability, stability over a wide pH range, greater solubility, and biocompatibility. Moreover, the compounds produced by *Aspergillus* are classified as generally regarded as safe (GRAS) status, which can be safely used in the industry [6]. NPs fabricated by fungi have been used for different applications such as in medicine, as an anti-cancer drug, antibiotic, antifungal, antimicrobial, and antiviral agents [7], in diagnostic, bioimaging, biosensor, agricultural, and other industrial applications [8]. A new term "Myconanotechnology", was proposed by Rai and co-workers [9] to highlight the research on fungi in the production of NPs and their role in the nanotechnology research.

Industrial biotechnology processes demonstrate a significant reduction of greenhouse gas emissions using renewable resources. The process is environment friendly and do not result in the accumulation of toxic compounds in the ecosystem. In industrial biotechnology, biomass input is used under the process of biological agents like metabolites and biomolecules to create a wide spectrum of products. There is a worldwide interest to enable the production of different NPs on biotechnological lines because of their eco-friendly nature, less energy-intensive, ease of execution, and ability to modify biological agents, and products [10].

In the present chapter, we are going to focus on the need for large-scale productions of NPs by biological methods in general and by *Aspergillus* spp. in particular. Different NPs fabricated by the *Aspergillus* spp., their advantages over the other methods, details of the mechanistic aspects of nanoparticle production, and various applications of fabricated NPs. Toxicity concerns of the large-scale production of NPs will also be discussed.

2. Diversity of *aspergillus* spp. for the synthesis of different nanomaterials

More than 6400 different biologically active substances have been reported from filamentous fungi which have potential bioactivities and different applications [11]. As these fungi have greater tolerance to high metal ion concentration and have the ability to internalize and bio accumulate metal ions they can be used for metal ion reduction and stabilization in nanomaterial synthesis [12–16]. A huge range of fungi is shown to have the ability to synthesize NPs. Out of which *Aspergillus* is one of the major contributors in the mycosynthesized (fungus mediated synthesis) NMs with various biological activities. A huge range of *Aspergillus* spp. have been reported to synthesize different NMs including metal and metal oxide NPs. The cell-free extracts, as well as supernatant of fermented medium, can also be used for the synthesis of NPs [14, 15, 17]. The following **Table 1**, summarizes the *Aspergillus* spp. and the respective NMs synthesized by them.

Aspergillus spp.	Nanomaterial synthesized	Reference
Aspergillus tubingiensis	Silver	[18]
Aspergillus niger IPT856	Silver	[19]
Aspergillus oryzae	Silver	[20]
Aspergillus flavus	Silver	[21]
Aspergillus versicolor	Silver	[22]
Aspergillus terreus	Silver	[23]
Aspergillus versicolor	Silver	[24]
Aspergillus oryzae (MTCC no. 1846)	Silver	[25, 26]
Aspergillus fumigatus BTCB10	Silver	[27, 28]
Aspergillus niger	Silver	[14, 15]
Aspergillus tamari *Aspergillus niger*	Silver	[29]
Aspergillus flavus	Silver	[30]
Aspergillus terreus	Silver	[31, 32]
Aspergillus flavus	Silver	[33]
Aspergillus fumigates	Silver	[34]
Aspergillus oryzae var. *wiridis*	Silver	[35]
Aspergillus flavus	Silver	[36]
Aspergillus fumigatus DSM819	Silver	[37]
Aspergillus niger	Silver	[17]
Aspergillus niger	Silver	[38]
Aspergillus niger	Silver	[39]
Aspergillus fumigatus	Silver	[40]
Aspergillus flavus	Silver	[41]
Aspergillus clavatus	Silver	[42]
Aspergillus flavus NJP08	Silver	[43]
Aspergillus terreus CZR-1	Silver	[44]
Aspergillus terreus	Silver and Gold	[45]
Aspergillus sydowii	Gold	[46]
Aspergillus niger	Gold	[47]

Aspergillus spp.	Nanomaterial synthesized	Reference
Aspergillus niger	Gold	[48]
Aspergillus terreus IF0	Gold	[49]
Aspergillus niger	Gold	[50]
Aspergillus clavatus	Gold	[51]
Aspergillus flavus	TiO_2	[52]
Aspergillus flavus TFR7	TiO_2	[53]
Aspergillus niger	ZnO	[54]
Aspergillus fumigatus	ZnO	[55, 56]
Aspergillus oryzae	$FeCl_3$	[57]
Aspergillus tubingensis	$Ca_3P_2O_8$	[58]
Aspergillus versicolor mycelia	Hg	[59]
Aspergillus aureoterreus Samson et al. AUMC 13006	CuO	[60]
Aspergillus carneus Blochwitz AUMC 13007	CuO	[60]
Aspergillus flavus var. *columnaris* Raper and Fennell AUMC 13012	CuO	[60]
Aspergillus fumigatus Fresenius AUMC 13024	CuO	[60]
Aspergillus sydowii (Bainier and Sartory) Thom and Church	CuO	[60]
Aspergillus terreus Thom *AUMC 13019*	CuO	[60]

Table 1.
Various aspergillus *spp. and respective nanomaterials synthesized by them.*

The cell-free extracts of *Aspergillus* spp. are challenged against the precursor salt for direct synthesis of NPs. But *Aspergillus* extract fermented lupin was reported by Mosallam et al., [61], for the biological synthesis of selenium NPs in the presence of gamma radiation. Balakumaran et al. [45] reported the various strains of *Aspergillus* isolated from Kolli hills and Yercaud hills, South India, and identified their ability to synthesize extracellular gold and silver NPs. Out of all the screened isolates, *A. terreus* showed the most stable nanoparticle synthesis. Vala [46] reported the synthesis of gold NPs by marine-derived fungus *Aspergillus sydowii* [62]. The intracellular synthesis of gold NPs by *Ammophilus fumigatus* has been reported by Bathrinarayan et al., [63, 64]. In one of the studies on the experimental rat model have demonstrated the wound healing ability of *A. niger* mediated silver NPs [65]. Ghareib et al., [60] isolated *Aspergillus* strains from Egyptian soil and reported their biomass and culture supernatant mediated synthesis of copper oxide (CuO) NPs. Biogenic zinc oxide NPs are synthesized from the cell-free fungal filtrate of *A. niger,* which has antimicrobial and dye degradation ability [54].

All these various types of NPs synthesized using different isolates and strains of *Aspergillus* are shown to have distinguishing bioactivities, numerous functions, and applications in various fields [14, 15, 66]. These fungus-mediated metal and metal oxide NPs are synthesized intra or extracellularly and are reported to appear in various shapes and sizes [67].

3. Advantages of nanomaterial production by *aspergillus* spp

The green chemistry approach highlights the usage of microorganisms which offers a cheaper, lighter, reliable, nontoxic, and eco-friendly process [68, 69]. Fungi secrete a higher amount of proteins owing to significantly higher productivity of NPs [70] which effectively proved a potential source for the extracellular synthesis of different NPs without using harmful toxic chemicals. The advantages made fungi more suitable for large-scale production and easy downstream processing, also economic [70, 71]. Besides, enzyme nitrate reductase is found to be responsible for the synthesis of NPs in fungi [68, 69]. Biofabrication of NPs using fungi (eukaryotic organism) has several advantages over the prokaryotic mediated approach for reproducibility of nanosized materials. Also include ease to multiplication, grow, handling, and rest of downstream process for this top-down approach of nanobio-synthesis through nano factories [72, 73]. Tarafdar et al., [74] observed rapid, low cost, and eco-friendly iron nanoparticle fabrication by using the fungi *Aspergillus oryzae* TFR9. This study evaluated the morphological and elemental characterization of the biosynthesized iron NPs [74].

Zielonka et al., [75] demonstrated fungi are almost ideal biocatalysts for NPs biosynthesis. In contrast to bacteria, as they are well-known for producing greater amounts of biologically active substances that make the fungus more appropriate for large-scale production [31, 32]. Moreover, fungal biomass can resist flow pressure, agitation, and harsh conditions in chambers such as bioreactors. Also, they exude extracellular reductive proteins which can be used in subsequent process steps. However, the fungal cell is deprived of unessential cellular components since NPs are accelerated outside the cell and can be immediately used in manifold ways without pre-treatment [76]. There are a large number of fungi, which can efficiently synthesize silver NPs, such as *Aspergillus clavatus,* and have many biomedical applications [77].

Here we highlighted the advantages of NMs produced by using *Aspergillus* spp. The high-scale production of NPs from fungi has wide applications in protein engineering, synthetic biology, and downstream processing (**Figure 1**). For large-scale production, fungi can be effectively employed. Gade et al., [14, 15] studied extracellular biosynthesis of AgNPs by *Aspergillus niger* isolated from soil. The nitrate-dependent reductase enzyme reduced the silver ions and a shuttle quinone extracellular process. The reduction of silver ions was an extracellular process.

AgNPs released silver ions in the fungal cell, which increased its antifungal function. AgNPs synthesized by using *A. terreus* HA1N sp. There is a large number of fungi, which can efficiently synthesize silver NPs, such as *Aspergillus clavatus* (*A. clavatus*) or ZnO NPs are produced by *Aspergillus terrus* [57]. The effect of prepared zinc oxide NPs on the growth and mycotoxin production by mycotoxigenic molds was evaluated which was concentration-dependent. The levels of produced mycotoxins were decreased when the concentration of ZnONPs increased [78]. Bathrinarayanan et al., [63, 64] produced gold NPs by *Aspergillus fumigatus*. It was found to be stable, spherical, and had irregular morphologies which were confirmed by SEM analysis.

El-Desouky et al., [79] demonstrated the synthesis AgNPs by an eco-friendly and low-cost method using the fungi *Aspergillus terreus* HA1N. It is an alternative to chemical procedures which require drastic experimental conditions emitting toxic chemical byproducts. The AgNPs are widely used as a novel therapeutic agent as antibacterial, antifungal, antiviral, anti-inflammatory, and anti-cancer agents [80, 81]. The AgNPs synthesized by *Aspergillus* ssp. present potential advantages such as fast growth rate, the rapid capacity of metallic ion reduction, nanoparticle stabilization, and facile and economical biomass handling. Moreover, these fungi have significantly higher productivity when used in nanoparticle biosynthesis due to their higher protein secretion [82]. Husain et al., [83] demonstrated the immobilization of *Aspergillus oryzae* β-galactosidase on native ZnO and zinc oxide NPs (ZnO-NP) by using a simple adsorption mechanism. The ZnO has wide

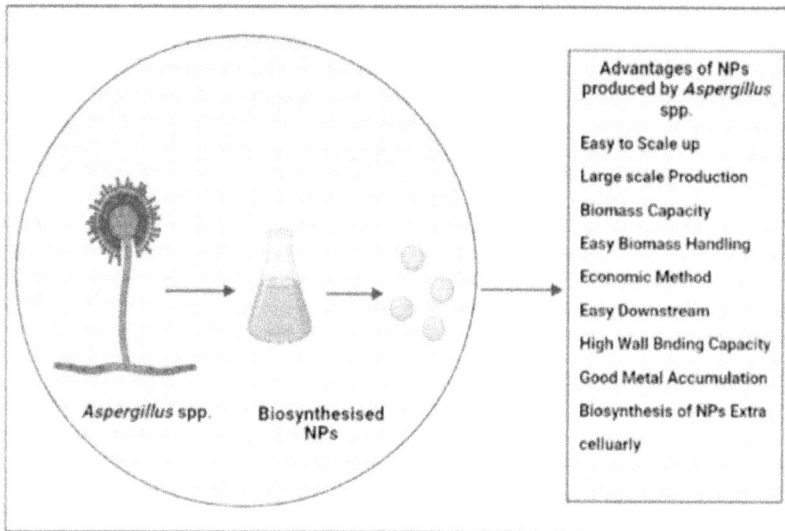

Figure 1.
Advantages of nanoparticles produced by aspergillus *spp.*

applications. In addition to this easy production, improved stability against various denaturants, and excellent reusability, ZnO-NP bound β galactosidase. There are many applications in constructing enzyme-based analytical devices for clinical, environmental, and food technology. *Aspergillus niger* showed effective fabrication of AgNPs [84]. The mycotoxin produced by mycotoxigenic fungi such as *Aspergillus* sp. Food toxin can be detected by nano-based biosensor. The functionalized NMs are used as catalytic tools, immobilization platforms, or optical or electroactive labels to improve the biosensing performance to obtain higher sensitivity, stability, and selectivity. Recently, these nano biosystems are also bringing advantages in terms of the design of novel food toxin detection strategies [85].

4. Mechanistic aspect of nanomaterial synthesis by *aspergillus* spp

It is well-identified that biological systems can fabricate the number of metallic and non-metallic nanoparticles. Synthesis of nanoparticles can be achieved at low cost by biological system especially from the fungal system at low pH, temperature, and salt concentration. Various studies have been proved that fungus-like *Fusarium* [7], *Phoma* [4], *Aspergillus* [86], and many more were found to be excellent factory for synthesis different types of nanoparticles. Every fungal species has unique biomolecule contents, which play a crucial role in the synthesis of nanoparticles. Due to this convolution still, an exact mechanism for nanoparticles synthesis from specific fungal species is yet to be revealed.

Even though, various studies have been initiated to understand the mechanism for the synthesis of nanoparticles from *Aspergillus* spp. Jain and co-workers [43] proposed two-step mechanism for silver nanoparticles synthesis from *Aspergillus flavus* NJP08. In the first step, 32 kDa reductase protein secreted by fungus might be responsible for the synthesis, and in the next step 35 kDa protein is responsible to provide stability to silver nanoparticles. In one of the study by Phanjom and Ahemad [86] proposed that the nitrate reductase enzyme secreted by *Aspergillus oryzae* (MTCC No. 1846) is responsible for the conversion of Ag + to Ag0. Selenium

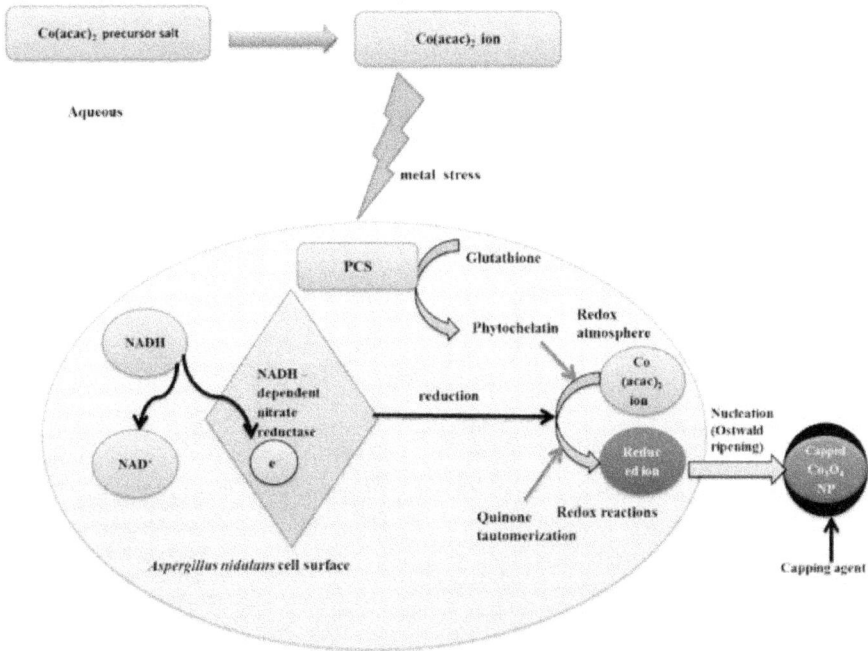

Figure 2.
Possible mechanism for the biosynthesis of Co$_3$O$_4$ nanoparticles in A. nidulans *[87].*

nanoparticles are also proved to be synthesized by the aqueous extract of fermented Lupin using *Aspergillus oryzae* and nucleation by gamma-ray (30.0 kGy) [61]. The authors confirmed that due to unique characteristics and novel biosynthesis method, selenium nanoparticles could be a good green antimicrobial candidate in biomedicine, cosmetics, and pharmaceutics. Endophytic fungi *Aspergillus nidulans* also produced cobalt oxide nanoparticles through the detoxification mechanism. In the synthesis, nitrate reductase along with electron shuttling compounds and other peptides are responsible for the reduction and synthesis [87]. Pavani et al. [88] reported the possible reductase or cytochrome base synthesis of lead nanoparticles from *Aspergillus* species. In the first step, the mechanism stated the trapping of lead ions on the fungal cell wall through electrostatic attraction. In the next step, these ions get entered into the cell and might get reduced by enzymes existing in the cell wall and inside the cell wall. In one of the study conducted by Li et al., [31, 32] reported the stabilized nanoparticles synthesis through reducing agent nicotin-amide adenine dinucleotide (NADH) present in the *Aspergillus terreus* (**Figure 2**).

5. Applications of nanomaterials synthesized using *aspergillus* spp

The numerous NMs have been synthesized by *Aspergillus* spp. and applied in various fields, for instance, Ag, Cu, Fe, Fe$_3$O$_4$, and ZnO NMs are some of them [55, 56, 72, 89–95]. Nanotechnology platform finds application of NMs in almost in each and every field such as agriculture, environmental science, health sciences, portable water treatment large/small scale plants, industrial separation, catalyst, electronics, energy storage, and energy regeneration [96–99]. The applications of NMs synthesized using *Aspergillus* spp. in various fields are shown in the graphical representation of **Figure 3**.

- Nanomedicine
- Pharmaceutical formulations
- Biotechnological and biomedical applications
- Sensors
- Drug and gene delivery systems

- Increasing crop yield
- Fertilizers
- Plant growth promoters
- Soil improvers
- Nanoencapsulation for slow release of agrochemicals

Medicine and Pharmacy

Agriculture

Applications of NMs by Aspergillus spps.

Environment management

Food security and Animal Industry

- Water disinfectant
- Antimicrobial surface coatings
- Renewable energy
- Environmental sensing

- Food packaging
- Nanosensors
- Antimicrobial agent
- Animal feed supplement

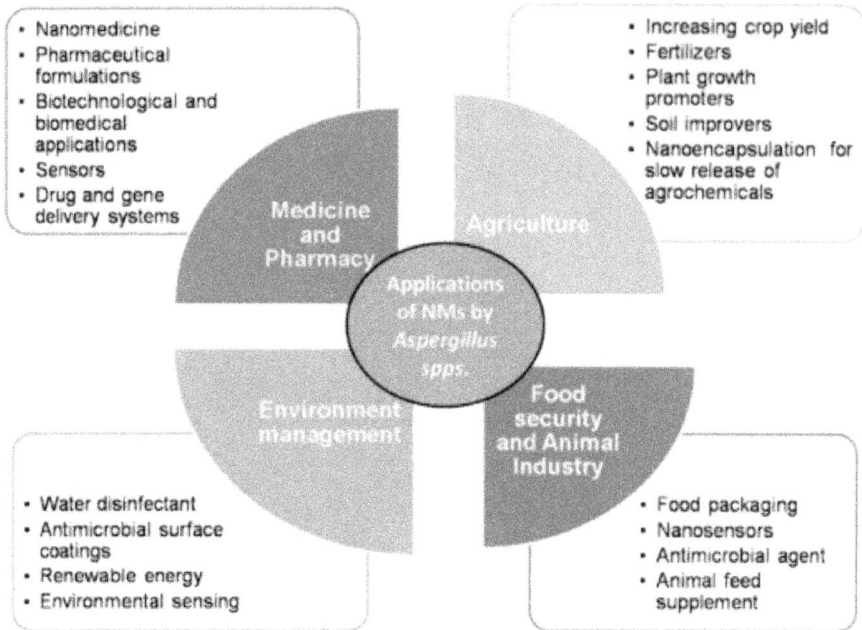

Figure 3.
Graphical representation of different applications of NMs synthesized using aspergillus *spp.*

NMs synthesized by *Aspergillus* species have tremendous applications broadly in the areas like agriculture, food security and animal industry, environmental management, and medicine and pharmacy are some of them. The detailed description is given in the sections below.

5.1 Applications in agriculture

In recent years, the nanotechnological advances in the field of agriculture have been increasing as the application of various NMs in the development of nano-based products like nanofertilizers for increasing crop yield and soil improvement, for plant growth promotion, nanopesticides, nanofungicides, nanoencapsulation for slow release of agrochemicals, and more in which NMs plays a vital role. The application of NPs as agrochemicals has become more common as technological advances make their production more economical for employment in the agriculture sector. For the potential application of NPs in plant disease control primarily included the information about the antimicrobial activity of different nano-size compounds against phytopathogens and the development of better application strategies to enhance the efficacy of disease suppression [100]. The antimicrobial activity of *Aspergillus* spp. synthesized NMs have been reported by many researchers. Silver NPs have been reported as most effective against phytopathogenic fungi, *Magnaporthe grisea* and *Bipolaris sorokiniana, in-vitro* [101], *Alternaria solani* and *Erwinia cartovora* pv. *cartovora* [102], and *Alternaria* blight as well as *Phytophthora* blight [103]. Elgorban et al., [24] demonstrated the silver NPs synthesis using fungus *Aspergillus versicolor* and evaluate its antifungal activity against *S. sclerotiorum* and *Botrytis cinerea* in strawberry plants. The silver NPs showed the concentration-dependent activity towards both the tested organisms but showed the greatest effect against *B. cinerea*. In another study, Ismail et al., [104] evaluated the combined effect of silver and selenium NPs against fungus *A. solani* that causing early blight disease of potato. The fungus isolated from leaf spot was identified by microscopy and treated with NPs suspension, which showed the formation of pits and pores. Therefore, the authors concluded that the myco-synthesized AgNPs were able to penetrate and distribute throughout the fungal cell area and interact with the components, and cause cell death. Silver/chitosan nanoformulations (NFs) were applied against various seed-borne plant pathogens, particularly seed-borne disease-causing fungi, isolated from chickpea seeds [105]. These studies reveal the possibilities of NPs application as an antifungal agent, alternative to the fungicide for controlling plant pathogens.

Nanoformulations of copper-chitosan (Cu/Ch) has been prepared as an anti-fungal agent against *A. solani* that causing early blight disease of tomato (*Solanum lypersicum* Mill). These NPs caused mycelia growth inhibition and spore germina-tion in *A. solani* and *F. oxysporum*, respectively, *in-vitro* model [106]. Recently, Shende et al., [94] synthesized the CuNPs using *Aspergillus flavus* and tested its activity against selected fungal plant pathogens namely *Aspergillus niger, Fusarium oxysporum,* and *Alternaria alternata*, which reveals significant antifungal activity. The study suggested the application of CuNPs as an effective fungicide for sustain-able agriculture. ZnO NPs have also been investigated as an effective fungicidal agent against plant pathogens. ZnO NPs have many advantages over silver NPs for fungal pathogen control efforts [107]. He et al., [108] evaluated the antifungal effect of ZnO NPs and their mode of action against two post-harvest pathogenic fungi *viz. B. cinerea* and *Penicillium expansum*. Different concentrations of NPs, when applied to fungal hyphae demonstrated cell wall damage and collapse fun-gal hyphae. Raliya and Tarafdar [55, 56] reported the synthesis of ZnO NPs by *Aspergillus fumigatus* TFR-8, with the size range between 1.2 ~ 6.8 nm and Oblate spherical and hexagonal shape and evaluated its effect on phosphorous-mobilizing enzyme secretion and gum contents in cluster bean (*Cyamopsis tetragonoloba* L.). The antibacterial potential of photocatalytic nanoscale titanium dioxide (TiO_2), nanoscale TiO_2 doped with zinc (TiO_2/Zn; Agri-Titan), and nanoscale TiO_2 doped (incorporation of other materials into the structure of TiO_2) with a silver (TiO_2/Ag) has been evaluated against *Xanthomonas perforans*, bacteria causing bacterial spot disease in tomato [109]. Shenashen et al., [110] synthesized and characterized the mesoporous alumina sphere (MAS) NPs and evaluated their biological activity against *F. oxysporum*, that causing root rot disease in tomatoes, in comparison with the recommended fungicide tolclofomethyl, under laboratory and green house conditions. The authors reported cell death because of the entry of NPs in fungal cells due to disruption of the cell membrane and malformation of hyphae.

5.2 Applications in food security and animal industry

The application of NMs in the food security and animal industry is attending the great interest of the scientific community in recent years. Food security is usually the preparation, treatment, and storage of food products in which the food-borne pathogens or illness will not going to cause any damage or spoilage to the product [96, 97, 111]. Food insecurity, like illegal additives, pathogens, pesticide residues, allergens, and other unsafe factors, those are not only seriously affects human health, but also limit the rapid development of food industries to a certain extent [112–114]. The identification and quantitative analysis of bacteria is a very impor-tant and crucial issue in food safety. Conventional practices require long culture time, highly skilled operators, or specific recognition elements of each type of bacteria [113]. For this purpose, the analytical methods or equipments that meet the requirement of modern detection of various hazardous substances present in the foods for example packaging materials, sensors, and food containers coated with

NPs are develop using NMs. The novel nano-based food packaging materials have the unique characteristics involving oxygen scavengers, antimicrobial potential, and barriers to gas or moisture, and many other. In view of these multiple benefits of nanopackaging, its application in the pathogens detection, antimicrobials, allergens and contaminants, UV-protecting activity, high gas barrier plastics, etc. are some important areas of research [115]. The use of such NMs in food packaging enhances the shelf life of food devoid of undesirable alteration in its quality.

The application of smart packaging systems has increased tremendously in animal industries the muscle-based food products such as meat, chicken, etc. that are prone to contamination. The packaging of meat and muscle products suppress the spoilage, enhance the tenderness by allowing enzymatic activity, avoid contamination, retain the cherry red color in red meats and reduce the loss in its weight [116]. Plastic food packaging is one of the most important areas of research that employ nanotechnology to make stronger and lighter packaging materials and also enhances its performance. Besides this, NMs with strong antimicrobial properties such as Ag and TiO_2 NPs could be used in the packaging of foods to prevent spoilage [117]. Additionally, the application of NPs of clay in food packaging helps to control the entry of carbon dioxide, oxygen, and moisture towards food materials, thus preventing food spoilage.

Nowadays, more researchers have been paying attention to the development of nanosensors, which are being added in plastic packaging to spot the gases released from spoiled food. In the food spoilage or contamination condition, the packaging material will alert the consumer by detecting toxins, microbial contamination, and pesticides in food products, based on flavor production and color changing [118]. Moreover, plastic films entrenched with silicate NPs are being developed to maintain food fresh for a longer period. In this case, NPs play a vital role in dropping the oxygen flow and also facilitate to impede the moisture seeping out from the package. In animal industries, *Aspergillus* spp. synthesized NMs such as ZnO NPs are used as antimicrobial agents. For instance, *Aspergillus fumigatus* JCF and *Aspergillus niger* synthesized ZnO NPs with 60 ~ 80 nm and 61 ± 0.65 nm size, respectively and Spherical shape demonstrated the antimicrobial potential [54, 93]. The antifungal activity was observed in the ZnO NPs, which were synthesized by *Aspergillus terreus* [90]. Recently, Mausa et al., [119] reported the mycosynthesis of various NMs *viz.* Co_3O_4, CuO, Fe_3O_4, NiO, and ZnO NPs by endophytic fungus *Aspergillus terreus*, and studied its antimicrobial and antioxidant activities, which leads to their application in different fields.

5.3 Applications in healthcare, medicine, and pharmacy

In medicine and pharmacy, NMs have been successfully applied due to their high surface area that is able to adsorbed or conjugate with an extensive variety of therapeutic and diagnostic agents such as drugs, vaccines, genes, antibodies, and biosensors. In recent years, antibiotic resistance is an emerging major global health problem and novel antimicrobial formulations are essentially needed to fight against these drug-resistant microbes, therefore nano-based medicine as antimicrobial agents have gained considerable attention in the field of microbial drug resistance [119, 120]. Hence, the NPs synthesized by Mousa et al., [119] using the endophytic fungus *Aspergillus terreus* were studied to discover their efficacy against different multi-drug-resistant bacterial strains as well as some human and plant pathogenic fungi. The authors have reported the broad-spectrum antimicrobial action of all the mycosynthesized NPs, where the bacterial and fungal strains were inhibited. Furthermore, Co_3O_4 NPs among the five types of mycosynthesized NPs exhibited the strongest antimicrobial potential against the tested pathogens. There are very few reports on the antimicrobial activity of Co_3O_4 NPs and only very less reported their antibacterial potential only [121]. Meanwhile, previous reports have observed the antimicrobial activity of CuO, Fe_3O_4, NiO, and ZnO NPs [122–125].

There are several reports on the synthesis and antimicrobial applications of *Aspergillus* spp. synthesized NMs. Netala et al., [22] demonstrated the antibacterial activity of Ag NPs synthesized by *Aspergillus versicolor* against *Staphylococcus aureus*, *Streptococcus pneumonia*, *Pseudomonas aeruginosa*, and *Klebsiella pneumoniae* at concentration 1 mg/mL. In another study, *Aspergillus terreus*-mediated synthesis of Ag NPs, showed antibacterial activity against *Salmonella typhi*, *S. aureus*, and *Escherichia coli* [23]. The synergistic effect with Ag NPs synthesized by *Aspergillus flavus* and conventional antibiotics against multi-drug-resistant bacteria such as *Bacillus spp.*, *Micrococus luteus*, *S. aureus*, *Enterococcus faecalis*, *E. coli*, *P. aeruginosa*, *Acinetobacter baumanii*, and *K. pneumoniae* at concentration 100 ppm [21]. Rodrigues et al., [18] reported the synthesis of Ag NPs using *Aspergillus tubingiensis* and demonstrated the antimicrobial activity against *Candida* sp. and *P. aeruginosa* at concentrations 0.11-1.75 μg/mL and 0.28 μg/mL respectively. Ag NPs synthesized by *Aspergillus oryzae* revealed the antifungal effect against *Trichophyton rubrum* at concentration > 7.5 μg/mL [20]. Another strain of *Aspergillus oryzae* (MTCC no. 1846) synthesized Ag NPs using 1 mM $AgNO_3$, produced 7-27 nm-sized spherical particles, which showed the antibacterial effects [25, 26]. Ottoni et al., [19] reported the synthesis of Ag NPs by *Aspergillus niger* IPT856 and its antibacterial activity against *E. coli*, *S. aureus*, and *P. aeruginosa*. In another study, *Aspergillus fumigatus* BTCB10 synthesized Ag NPs with a spherical shape, which demonstrated the antibacterial and cytotoxic effects [27, 28]. The ZnO NPs as a dietary supplement in the animals gives health benefits, which improves the quality of egg in poultry, help in wound healing, act as an antioxidant, improve growth performance, hormone production, bone formation, immune system, a cofactor for enzymatic process, and reproduction system [126]. The antibacterial and antifungal potential improves the health of the livestock. In another study, Farrag et al., [92] reported the synthesis of Ag NPs by *Aspergillus niger* isolated from soil by treatment with silver nitrate. AgNPs exhibited significant inhibition of *Allovahlkampfia spelaea* viability and growth of both trophozoites and cysts, with a reduction of amoebic cytotoxic activity in host cells that suggested the Ag NPs possibly will give a promising future for the treatment of *Allovahlkampfia spelaea* infections in humans.

5.4 Applications in environmental management

NMs offer a unique platform for the purification of water contaminated with pollutants namely organics, metal ions, biological contaminants, and arsenic from the water because of the high surface area of nanosorbents and their ability of chemical modification as well as easier regeneration [127–130]. Chatterjee et al., [91] reported the synthesis of superparamagnetic iron oxide NPs (IONPs) (Fe_3O_4) of 20-40 nm size by manglicolous (mangrove) fungus *Aspergillus niger* BSC-1 and employed for the removal of hexavalent chromium from aqueous solution. Therefore, suggested the utilization of mycosynthesized IONPs could be employed for the heavy metal remediation from contaminated wastewater. The enzymatic bioremediation of textile industry wastewater containing direct green or reactive red azo dye by utilization of enzymes immobilized onto magnetic NPs for the improvement of industrial and environmental applications have been reported by Darwesh et al., [131]. Different types of magnetic NPs have been used to remove heavy metal ions from industrial wastewater [132].

In another study, the Au NPs was synthesized by *A. niger* that was found to be very effective against the mosquito larvae. The AuNPs were tested using the larvae of three mosquito species viz. *Anopheles stephensi, Culex quinquefasciatus*, and *Aedes aegypti*. Among them, it has been observed that the effect of Au NPs was found to be significant against *C. quinquefasciatus* larvae than the *A. stephensi* and *A. aegypti* larvae. All larval instars of *C. quinquefasciatus* showed 100% mortality after 48 hours of exposure to the Au NPs synthesized by *A. niger* [50]. In conclusion, the authors suggest that the application of mycosynthesized Au NPs by *A. niger* could be the fast and environmentally friendly approach towards the control of mosquitoes than the currently available approaches. This may possibly lead to a novel potential strategy for vector control [50].

Other than this, nowadays NMs could be applied in antimicrobial surface coatings, environmental sensing, renewable energy, and many other environmental applications.

6. Toxicity aspects of nanomaterials

Assessment of toxicity of synthesized NPs is the critical step for ensuring their safe and sustainable applications. Hence, toxicity evaluation of all the newly synthesized nanoparticle must be considered before their industrial applications. As far as the comparison of biosynthesized NPs with NMs synthesized by other methods especially the chemical method is concerned, the biosynthesized NPs seems to be biocompatible [133]. For instance, the green synthesized NPs were found to enhance the plant seedling growth, yield and quality, suggesting the biocompatibility of biosynthesized NPs as compared to the chemical synthesis NPs [134]. In contrast, few studies have shown the toxicity of biosynthesized or green synthesized NPs. Sulaiman et al., [135] have synthesized silver NPs (AgNPs) by using *Aspergillus flavus* and investigated its cytotoxic effect on HL-60, a human promyeloid leukemia cells. The study reported the dose-dependent toxicity of AgNPs at concentration of 5 and 10 μg/ml. The study claimed that although AgNPs have a toxic effect on normal cells but they also have the potential to act as a potential anticancer agent. However, the toxicity of AgNPs was found to be higher than silver nitrate solution. The said toxic effect was claimed to be due to the physicochemical interaction of silver atoms of AgNPs with the functional groups of intracellular proteins, nitrogen bases, and phosphate groups of DNA. The AgNPs may induce the accumulation of reactive oxygen species (ROS) leading to cellular apoptosis. Such effect will be helpful for the anti-cancer, antiproliferative, and antiangiogenic effects *in vitro*. Othman et al., [136] have synthesized AgNPs by using *A. terreus* and studied its antitumor activity against Human Caucasian breast adenocarcinoma (MCF7). It was found to inhibit the growth in dose-dependent manner with IC_{50} value of 46.7 μg/ml. Similarly, ZnO NPs synthesized by using the culture filtrate of *A. terreus* have been reported to be cytotoxic to HeLa cells. It was shown to induce apoptosis by inhibiting the production of cellular superoxide dismutase, catalase, glutathione peroxidase levels and inducing the accumulation of ROS, and reduction in mitochondrial membrane potential. Moreover, further investigation found it to induce oxidative damage via down-regulating expression of p53, Bax, Caspase-3, Caspase-9, and up-regulating Cytochrome-C expression [137]. The nanoparticle when come in contact with the target cell membrane, gets accumulates at the cell surface and induces pore formation causing the leakage of cytoplasmic material outside the cell. The NPs entered in the cell can interact with intracellular protein and DNA and thus disturbing the cell regulation [138]. The discussed mechanism of nanoparticle toxicity, in general, is represented in **Figure 4**. In general, toxicity studies are performed on human and animal cell line and plants. But there is also a need to give attention to the toxicity studies on microbes that are directly

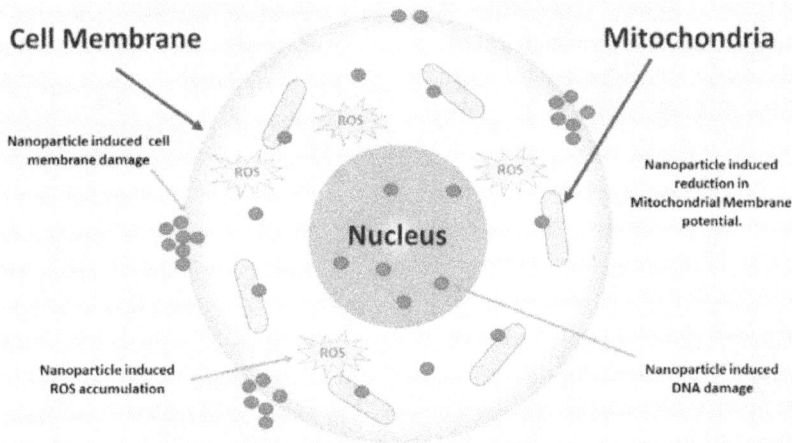

Figure 4.
Mechanism of cytotoxicity of mycosynthesized nanoparticles.

or indirectly beneficial to humans. Gupta et al., [139] have explored the toxicity of AgNPs synthesized by different fungi. The study found the mycosynthesized AgNPs to show toxicity to soil beneficial bacteria *Pseudomonas putida* KT2440 at the concentration of 0.4 µg/ml. Mycosynthesized Selenium nanoparticles (Se-NPs) were found to alter Wi-38, a normal lung fibroblast cells. At the IC50 value of 461 ppm, it exerted the loss of typical cell shape, granulation, loss of monolayer, and shrinking or rounding of cells [140].

Considering all of these observations from various studies it is suggested that before the actual application of any biosynthesized nanoparticle there is a need to undertake the toxicity studies and then make their use at biocompatible dose. For example, *A. terreus* strain AF-1 was exploited for the synthesis of CuO NPs which were integrated in cotton fabric. The said nanoparticle was used at a safe dose making it applicable for desired purpose [141]. Although mycosynthesized NPs are quite biocompatible as compared to NPs synthesized by other methods, it is recommended to verify the toxic effects of each type of NPs and choose their safe dose so as to make them safer for any kind of application.

7. Conclusion

Nanomaterials as the structures fabricated in the nanoscale have gained increasing attention for diagnostic and therapeutic purposes especially for those produced in a green safe approach by using fungi and other microorganisms. Among fungal species successfully used for this purpose, members of the genus Aspergillus are in the first line of investigation because of their huge diversity and capability to grow in abundance in laboratory conditions. Although there are many reports on the synthesis and biological activities of nanomaterials of different origins by fungi, little has been documented about important disciplines such as their mode of action and applications in medicine and industry. This chapter has highlighted the diversity of *Aspergillus* species and their advantages for nanomaterial production, mechanistic aspects of nanomaterial synthesis by selected Aspergilli, applications in healthcare, medicine, and pharmacy, role in environmental management as a unique platform for water decontamination, and finally, the cytotoxicity of introduced nanomaterials as a critical step for ensuring their safety and sustainability. Overall, these results further substantiate the importance and priority of *Aspergillus* species for nanomaterial production at an industrial scale in a safe and cost-effective manner which enables researchers to use them for diagnostic, detoxifying, and therapeutic purposes in industry, medicine, and agriculture.

Author details

Mahendra Rai[1*], Indarchand Gupta[2], Shital Bonde[1], Pramod Ingle[1], Sudhir Shende[1], Swapnil Gaikwad[3], Mehdi Razzaghi-Abyaneh[4] and Aniket Gade[1]

1 Nanobiotechnology Lab., Department of Biotechnology, Sant Gadge Baba Amravati University, Amravati, Maharashtra, India

2 Department of Biotechnology, Institute of Science, Aurangabad, Maharashtra, India

3 Microbial Diversity Research Center, Dr. D.Y. Patil Biotechnology and Bioinformatics Institute, Dr. D.Y. Patil Vidyapeeth (Deemed to be University), Pune, India

4 Department of Mycology, Pasteur Institute of Iran, Tehran, Iran

*Address all correspondence to: mahendrarai7@gmail.com; pmkrai@hotmail.com

IntechOpen

References

[1] Duran N, Marcato P, Duran M, Yadav A, Gade A, Rai M. Mechanistic aspects in the biogenic synthesis of extracellular metal nanoparticles by peptides, bacteria, fungi and plants. Applied Microbiology and Biotechnology. 2011;**90**:1609-1624

[2] Gade AK, Ingle AP, Whiteley C, Rai M. Mycogenic metal nanoparticles: Progress and applications. Biotechnology Letters. 2010;**32**(5):593-600

[3] Rai M, Bonde S, Golinska P, Trzcińska-Wencel J, Gade A, Abd-Elsalam K, et al. *Fusarium* as a novel fungus for the synthesis of nanoparticles: Mechanism and applications. Journal of Fungi. 2021;**7**(2):139

[4] Gade AK, Gaikwad SC, Duran N, Rai MK. Green synthesis of silver nanoparticles by *Phoma glomerata*. Micron. 2014;**59**:52-59

[5] Ward OP, Qin WM, Dhanjoon J, Ye J, Singh A. Physiology and biotechnology of *Aspergillus*. Advances in Applied Microbiology. 2005;**58**:1-75

[6] Gouka RJ, Punt PJ, van den Hondel CAMJJ. Efficient production of secreted proteins by *Aspergillus*: Progress, limitations and prospects. Applied Microbiology and Biotechnology. 1997;**47**:1-11

[7] Gaikwad S, Ingle A, Gade A, Rai M, Falanaga A, Incoronato N, et al. Antiviral activity of mycosynthesized silver nanoparticles against herpes simplex virus and human parainfluenza virus type 3. International Journal of Nanomedicine. 2013;**8**:4303-4314

[8] Moghaddam A, Namvar F, Moniri M, Tahir P, Azizi S, Mohamad R. Nanoparticles biosynthesized by fungi and yeast: a review of their preparation, properties, and medical applications. Molecules. 2015;**20**:16540-16565

[9] Rai M, Yadav A, Bridge P, Gade A. Myconanotechnology: A new and emerging science. In: Rai M, Bridge P, editors. Applied Mycology. 14th ed. NY: CAB international; 2009. pp. 258-267

[10] Birla SS, Gaikwad SC, Gade AK, Rai MK. Rapid synthesis of silver nanoparticles from *Fusarium oxysporum* by optimizing physicocultural conditions. The Scientific World Journal. 2013:796018. DOI: 10.1155/2013/796018

[11] Bérdy J. Bioactive microbial metabolites. The Journal of Antibiotics. 2005;**58**:1-26. DOI: 10.1038/ja.2005.1

[12] Ahluwalia V, Kumar J, Sisodia R, Shakil NA, Walia S. Green synthesis of silver nanoparticles by *Trichoderma harzianum* and their bioefficacy evaluation against *Staphylococcus aureus* and *Klebsiella pneumonia*. Industrial Crops and Products. 2014;**55**:202-206. DOI: 10.1016/j.indcrop.2014.01.026

[13] Azmath P, Baker S, Rakshith D, Satish S. Mycosynthesis of silver nanoparticles bearing antibacterial activity. The Saudi Pharmaceutical Journal. 2016;**24**:140-146. DOI: 10.1016/j.jsps.2015.01.008

[14] Gade AK, Bonde P, Ingle AP, Marcato PD, Duran N, Rai MK. Exploitation of *Aspergillus niger* for synthesis of silver nanoparticles. Journal of Biobased Materials and Bioenergy. 2008a;**2**:243-247

[15] Gade AK, Bonde P, Ingle AP, Marcato PD, Durán N, Rai MK. Exploitation of *Aspergillus niger* for synthesis of silver nanoparticles. Journal of Biobased Materials and Bioenergy. 2008b;**2**:243-247. DOI: 10.1166/jbmb.20 08.401

[16] Khan NT, Khan MJ, Jameel J, Jameel N, Rheman SUA. An overview:

Biological organisms that serves as nanofactories for metallic nanoparticles synthesis and fungi being the most appropriate. Bioceramics Development and Applications. 2017;7:101. DOI: 10.4172/2090-5025.1000101

[17] Gaikwad S, Bhosale A. Green synthesis of silver nanoparticles using *Aspergillus niger* and its efficacy against human pathogens. European Journal of Experimental Biology. 2012;2(5):1654-1658

[18] Rodrigues AG, Ping LY, Marcato PD, Alves OL, Silva MCP, Ruiz RC, et al. Biogenic antimicrobial silver nanoparticles produced by fungi. Applied Microbiology and Biotechnology. 2013a;97:775-782. DOI: 10.1007/s00253-012- 4209-7

[19] Ottoni CA, Simões MF, Fernandes S, Santos JG, Silva ES, Souza RFB, et al. Screening of filamentous fungi for antimicrobial silver nanoparticles synthesis. AMB Express. 2017a;7:31. DOI: 10.1186/s13568-017-0332-2

[20] Pereira L, Dias N, Carvalho J, Fernandes S, Santos C, Lima N. Synthesis, characterization and antifungal activity of chemically and fungal produced silver nanoparticles against *Trichophyton rubrum*. Journal of Applied Microbiology. 2014a;117: 1601-1613. DOI: 10.1111/jam.12652

[21] Naqvi SZH, Kiran U, Ali MI, Jamal A, Hameed A, Ahmed S, et al. Combined efficacy of biologically synthesized silver nanoparticles and different antibiotics against multidrug-resistant bacteria. International Journal of Nanomedicine. 2013a;8:3187-3195. DOI: 10.2147/IJN.S49284

[22] Netala VR, Bethu MS, Pushpalatah B, Baki VB, Aishwarya S, Rao JV, et al. Biogenesis of silver nanoparticles using endophytic fungus *Pestalotiopsis microspora* and evaluation of their antioxidant and anticancer

activities. International Journal of Nanomedicine. 2016a;11:5683-5696. DOI: 10.2147/IJN.S112857

[23] Rani R, Sharma D, Chaturvedi M, Yadav JP. Green synthesis, characterization and antibacterial activity of silver nanoparticles of endophytic fungi *Aspergillus terreus*. The Journal of Nanomedicine and Nanotechnology. 2017a;8:4. DOI: 10.4172/2157-7439.1000457

[24] Elgorban AM, Aref SM, Seham SM, Elhindi KM, Bahkali AH, Sayed SR, et al. Extracellular synthesis of silver nanoparticles using *Aspergillus versicolor* and evaluation of their activity on plant pathogenic fungi. Mycosphere. 2016a;7:844-852. DOI: 10.5943/ mycosphere/7/6/15

[25] Phanjom P, Ahmed G. Effect of different physicochemical conditions on the synthesis of silver nanoparticles using fungal cell filtrate of *Aspergillus oryzae* (MTCC No. 1846) and their antibacterial effects. Advances in Natural Sciences: Nanoscience and Nanotechnology. 2017a;8:1-13. DOI: 10.1088/2043-6254/aa92b

[26] Phanjom P, Ahmed G. Effect of different physicochemical conditions on the synthesis of silver nanoparticles using fungal cell filtrate of *Aspergillus oryzae* (MTCC No. 1846) and their antibacterial effects. Advances in Natural Sciences: Nanoscience and Nanotechnology. 2017b;8:1-13. DOI: 10.1088/2043-6254/aa92bc

[27] Shahzad A, Saeed H, Iqtedar M, Hussain SZ, Kaleem A, Abdullah R. Size-controlled production of silver nanoparticles by *Aspergillus fumigatus* BTCB10: likely antibacterial and cytotoxic effects. Journal of Nanomaterials. 2019a;5168698. DOI: 10.1155/2019/5168698

[28] Shahzad A, Saeed H, Iqtedar M, Hussain SZ, Kaleem A, Abdullah R.

Size-controlled production of silver anoparticles by *Aspergillus fumigatus* BTCB10: likely antibacterial and cytotoxic effects. Journal of Nanomaterials. 2019b;**5168698**. DOI: 10.1155/2019/5168698

[29] Devi LS, Joshi SR. Ultrastructures of silver nanoparticles biosynthesized using endophytic fungi. Journal of Microscopic Ultrastructures. 2015;**3**(1):29-37. DOI: 10.1016/j.jmau.2014.10.004

[30] Fatima F, Verma SR, Pathak N, Bajpai P. Extracellular mycosynthesis of silver nanoparticles and their microbicidal activity. Journal of Global Antimicrobial Resistance. 2016;**7**:88-92. DOI: 10.1016/j.jgar.2016.07.013

[31] Li G, He D, Qian Y, Guan B, Gao S, Cui Y, et al. Fungus-mediated green synthesis of silver nanoparticles using *Aspergillus terreus*. International Journal of Molecular Sciences. 2012a;**13**:466-476

[32] Li G, He D, Qian Y, Guan B, Gao S, Cui Y, et al. Fungus-mediated green synthesis of silver nanoparticles using *Aspergillus terreus*. International Journal of Molecular Sciences. 2012b;**13**(1):466-476. DOI: 10.3390/ijms13010466. Epub 2011 Dec 29

[33] Wilson A, Prabukumar S, Sathishkumar G, Sivaramakrishnan S. *Aspergillus flavus* mediated silver nanoparticles synthesis and evaluation of ITS antimicrobial activity against different human pathogens. International Journal of Applied Pharmaceutics. 2016;**8**(4):43-46

[34] Bhainsa KC, D'Souza SF. Extracellular biosynthesis of silver nanoparticles using the fungus *Aspergillus fumigates*. Colloids and Surfaces. B, Biointerfaces. 2006;**47**(2):160-164

[35] Binupriya AR, Sathishkumar M, Yun S. Myco-crystallization of silver

ions to nanosized particles by live and dead cell filtrates of *Aspergillus oryzae* var. Wiridis and its bactericidal activity toward *Staphylococcus aureus* KCCM 12256. Industrial and Engineering Chemistry Research. 2010;**49**:852-858

[36] Vidya P, Subramani G. Fungus mediated synthesis of silver nanoparticles using *Aspergillus flavus* and its antibacterial activity against selective food borne pathogens. Indo American Journal of Pharmaceutical Sciences. 2017;**4**(12):4627-4634. DOI: 10.5281/zenodo.1069710

[37] Othman AM, Elsayed MA, Al-Balakocy NG, Hassan MM, Elshafei AM. Biosynthesis and characterization of silver nanoparticles induced by fungal proteins and its application in different biological activities. Journal of Genetic Engineering and Biotechnology. 2019;**17**(1). DOI: 10.1186/s43141-019-0008-1

[38] Kathiresan K, Alikunhi NM, Pathmanaban S, Nabikhan A, Kandasamy S. Analysis of antimicrobial silver nanoparticles synthesized by coastal strains of *Escherichia coli* and *Aspergillusniger*. Canadian Journal of Microbiology. 2010;**56**:1050-1059

[39] Jaidev LR, Narasimha G. Fungal mediated biosynthesis of silver nanoparticles, characterization and antimicrobial activity. Colloids and Surfaces. B, Biointerfaces. 2010;**81**:430-433

[40] Alani F, Moo-Young M, Anderson W. Biosynthesis of silver nanoparticles by a new strain of *Streptomyces* sp. compared with *Aspergillus fumigatus*. World Journal of Microbiology and Biotechnology. 2012;**28**:1081-1086

[41] Vigneshwaran N, Ashtaputre NM, Varadarajan PV, Nachane RP, Paralikar KM, Balasubramanya RH.

Biological synthesis of silver nanoparticles using the fungus *Aspergillus flavus*. Materials Letters. 2007;**61**:1413-1418

[42] Saravanan M, Nanda A. Extracellular synthesis of silver bionanoparticles from *Aspergillus clavatus* and its antimicrobial activity against MRSA and MRSE. Colloids and Surfaces. B, Biointerfaces. 2010;**77**:214-218

[43] Jain N, Bhargava A, Majumdar S, Tarafdar JC, Panwar J. Extracellular biosynthesis and characterization of silver nanoparticles using *Aspergillus flavus* NJP08: A mechanism perspective. Nanoscale. 2011a;**3**:635-641

[44] Raliya R, Tarafdar JC. Novel approach for silver nanoparticle synthesis using *Aspergillus terreus* CZR-1: mechanism perspective. Journal of Bionanoscience. 2012;**6**:12-16

[45] Balakumaran MD, Ramachandran R, Balashanmugam P, Mukeshkumar DJ, Kalaichelvan PT. Mycosynthesis of silver and gold nanoparticles: Optimization, characterization and antimicrobial activity against human pathogens. Microbiological Research. 2016;**182**:8-20. ISSN 0944-5013. DOI: 10.1016/j.micres.2015.09.009

[46] Vala AK. Exploration on green synthesis of gold nanoparticles by a marine-derived fungus *Aspergillus sydowii*. Environmental Progress & Sustainable Energy. 2014. DOI: 10.1002/ep.11949

[47] Xie J, Lee JY, Wang DIC, Ting YP. High-yield synthesis of complex gold nanostructures in a fungal system. Journal of Physical Chemistry C. 2007;**111**:16858-16865

[48] Zhang X, He X, Wang K, Yang X. Different active biomolecules involved in biosynthesis of gold nanoparticles by three fungus species. Journal of Biomedical Nanotechnology. 2011;**7**:245-254. DOI: 10.1166/jbn.2011.1285

[49] Priyadarshini E, Pradhan N, Sukla LB, Panda PK. Controlled synthesis of gold nanoparticles using *Aspergillus terreus* IF0 and its antibacterial potential against gram negative pathogenic bacteria. Journal of Nanotechnology. 2014;**653198**. DOI: 10.1155/2014/653198

[50] Soni N, Prakash S. Synthesis of gold nanoparticles by the fungus *Aspergillus niger* and its efficacy against mosquito larvae. Reports in Parasitology. 2012;**2**:1-7

[51] Verma VC, Kharwar RN, Singh SK, Solanki R, Prakash S. Correction to biofabrication of anisotropic gold nanotriangles using extract of endophytic *Aspergillus clavatus* as a dual functional reductant and stabilizer. Nanoscale Research Letters. 2011;**6**. article 261

[52] Rajakumar G, Rahuman A, Roopan SM, Khanna VG, Elango G, Kamaraj C, et al. Fungus-mediated biosynthesis and characterization of TiO2 nanoparticles and their activity against pathogenic bacteria. Spectrochimica Acta. Part A, Molecular and Biomolecular Spectroscopy. 2012;**91**:23-29

[53] Raliya R, Biswas P, Tarafdar JC. TiO2 nanoparticle biosynthesis and its physiological effect on mung bean (*Vigna radiata* L.). Biotechnology Reports. 2015;**5**:22-26

[54] Kalpana VN, Kataru BAS, Sravani N, Vigneshwari T, Panneerselvam A, Devi RV. Biosynthesis of zinc oxide nanoparticles using culture filtrates of *Aspergillus niger*: Antimicrobial textiles and dye degradation studies. OpenNano. 2018a;**3**:48-55. DOI: 10.1016/j.onano.2018.06.001

[55] Raliya R, Tarafdar JC. ZnO nanoparticle biosynthesis and its effect on phosphorous mobilizing enzyme secretion and gum contents in Clusterbean (*Cyamopsis tetragonoloba* L.). Agricultural Research. 2013a;2:48-57

[56] Raliya R, Tarafdar JC. ZnO nanoparticle biosynthesis and its effect on phosphorous-mobilizing enzyme secretion and gum contents in cluster bean (*Cyamopsis tetragonoloba* L.). Agricultural Research. 2013c;2: 48-57. DOI: 10.1007/s40003-012-0049-z

[57] Raliya R. Rapid, low-cost, and ecofriendly approach her for iron nanoparticle synthesis using *Aspergillus oryzae* TFR9. Journal of Nanoparticles. 2013. DOI: 10.1155/2013/141274

[58] Tarafdar JC, Raliya R, Rathore I. Microbial synthesis of phosphorous nanoparticle from tri-calcium phosphate using *Aspergillus tubingensis* TFR-5. Journal of Bionanoscience. 2012;6:84-89

[59] Das S, Das A, Guha A. Adsorption behavior of mercury on functionalized *Aspergillus versicolor* mycelia: Atomic force microscopic study. Langmuir. 2008;25:360-366

[60] Ghareib M, Tahon MA, Abdallah WE, Tallima A. Green synthesis of copper oxide nanoparticles using some fungi isolated from the Egyptian soil. International Journal of Research in Pharmaceutical and Nano Sciences. 2018;7(4):119-128

[61] Mosallam FM, El-Sayyad GS, Fathy RM, El-Batal AI. Biomolecules-mediated synthesis of selenium nanoparticles using *Aspergillus oryzae* fermented Lupin extract and gamma radiation for hindering the growth of some multidrug-resistant bacteria and pathogenic fungi. Microbial Pathogenesis. 2018;122:108-116. ISSN 0882-4010. DOI: 10.1016/j.micpath.2018.06.013

[62] Kitching M, Ramani M, Marsili E. Fungal biosynthesis of gold nanoparticles: Mechanism and scale up. Microbial Biotechnology. 2015;8(6):904-917. DOI: 10.1111/1751-7915.12151

[63] Bathrinarayanan PV, Thangavelu T, Muthukumarasamy V, et al. Biological synthesis and characterization of intracellular gold nanoparticles using biomass of *Aspergillus fumigatus*. Bulletin of Materials Science. 2013a;36(7):1201-1205

[64] Bathrinarayanan VP, Thangavelu D, Muthukumarasamy VK, Munusamy C, Gurunathan B. Biological synthesis and characterization of intracellular gold nanoparticles using biomass of *Aspergillus fumigatus*. Bulletin of Materials Science. 2013b;36(7):1201-1205. DOI: 10.1007/s12034-013-0599-0

[65] Sundaramoorthi C, Kalaivani M, Mathews DM, Palanisamy S, Kalaiselvan V, Rajasekaran A. Biosynthesis of silver nanoparticles from *Aspergillus niger* and evaluation of its wound healing activity in experimental rat model. International Journal of PharmTech Research. 2009;1:1523-1529

[66] Sarsar V, Selwal MK, Selwal KK. Biogenic synthesis, optimisation and antibacterial efficacy of extracellular silver nanoparticles using novel fungal isolate *Aspergillus fumigatus* MA. IET Nanobiotechnology. 2016;10(4):215-221

[67] Siddiqi KS, Husen A. Fabrication of metal nanoparticles from fungi and metal salts: Scope and Application. Nanoscale Research Letters. 2016;11(1). DOI: 10.1186/s11671-016-1311-2

[68] Honary S, Barabadi H, Gharaei-Fathabad E, Naghibi F. Green synthesis of copper oxide nanoparticles using *Penicillium aurantiogriseum*, *Penicillium citrinum* and *Penicillium waksmanii*. Digest Journal of Nanomaterials and Biostructures. 2012a;7:999-1005

[69] Honary S, Gharaei-Fathabad E, Khorshidi Paji Z, Eslamifar M. A novel biological synthesis of gold nanoparticle by Enterobacteriaceae family. Tropical Journal of Pharmaceutical Research. 2012b;**11**:887-891

[70] Ahmad A, Mukherjee P, Senapati S, Mandal D, et al. Extracellular biosynthesis of silver nanoparticles using the fungus *Fusarium oxysporum*. Colloids and Surfaces. B, Biointerfaces. 2003;**28**:313-318

[71] Sagar G, Ashok B. Green synthesis of silver nanoparticles using *Aspergillus niger* and its efficacy against human pathogens. European Journal of Experimental Biology. 2012;**2**:1654-1658

[72] Jain N, Bhargava A, Tarafdar J, Singh S, Panwar J. A biomimetic approach towards synthesis of zinc oxide nanoparticles. Applied Microbiology and Biotechnology. 2013a;**97**(2):859-869

[73] Jain N, Bhargava A, Tarafdar JC, Singh SK, Panwar J. A biomimetic approach towards synthesis of zinc oxide nanoparticles. Applied Microbiology and Biotechnology. 2013b;**97**:859-869. DOI: 10.1007/s00253-012-3934-2

[74] Tarafdar C, Raliya R. Rapid, Low-Cost, and Ecofriendly Approach for Iron Nanoparticle Synthesis Using *Aspergillus oryzae* TFR9. Journal of Nanoparticles. 2013;**1-4**:141274

[75] Zielonka A, Klimek-Ochab M. Fungal synthesis of size-defined nanoparticles. Advances in Natural Sciences: Nanoscience and Nanotechnology. Advances in Natural Sciences: Nanoscience and Nanotechnology. 2017;**8**:043001

[76] Narayanan KB, Sakthivel N. Biological synthesis of metal nanoparticles by microbes. Advances in Colloid and Interface Science. 2010;**22**:156(1-2):1-13

[77] Verma VC, Kharwar RN, Gange AC. Biosynthesis of antimicrobial silver nanoparticles by the endophytic fungus *Aspergillus clavatus*. Nanomedicine. 2010;**5**(1):33-40

[78] Hassan AA, Howayda ME, Mahmoud HH. Effect of zinc oxide nanoparticles on the growth of mycotoxigenic mould. Study Chemical Process Technology. 2013:1-25

[79] El-Desouky TA, May MA, Naguib K. Effect of fenugreek seeds extracts on growth of aflatoxigenic fungus and aflatoxin B1 production. Journal of Applied Sciences Research. 2013;**9**:4418-4425

[80] Ammar HA, El-Desouky TA. Green synthesis of nanosilver particles by *Aspergillus terreus* HA1N and *Penicillium expansum* HA2N and its antifungal activity against mycotoxigenic fungi H.A.M. Journal of Applied Microbiology. 2016;**121**:89-100

[81] Otari SV, Patil RM, Ghosh SJ, Thorat ND, Pawar SH. Intracellular synthesis of silver nanoparticle by actinobacteria and its antimicrobial activity. Spectrochimica Acta. Part A, Molecular and Biomolecular Spectroscopy. 2015;**136**:1175-1180

[82] Mukherjee P, Ahmad A, Mandal D, Senapati S, et al. Fungus-mediated synthesis of silver nanoparticles and their immobilization in the mycelia matrix: A novel biological approach to nanoparticle synthesis. Nano Letters. 2001;**1**:515-519

[83] Husain Q, Ansari SA, Alam F, Azam A. Immobilization of *Aspergillus oryzae* β galactosidase on zinc oxide nanoparticles via simple adsorption mechanism. International Journal of Biological Macromolecules. 2011;**49**(1):37-43

[84] Hassan SA, Hanif E, Khan UH, Tanoli AK. Antifungal activity of silver

nanoparticles from *Aspergillus niger*. Pakistan Journal of Pharmaceutical Sciences. 2019;**1163-1166**

[85] Malhotra BD, Srivastava S, Ali MA, et al. Nanomaterial-Based Biosensors for Food Toxin Detection. Applied Biochemistry and Biotechnology. 2014;**174**:880-896

[86] Phanjom P, Ahmed G. Biosynthesis of silver nanoparticles by *Aspergillus oryzae* (MTCC No. 1846) and Its Characterizations. Nanoscience and Nanotechnology. 2015;**5**(1):14-21. DOI: 10.5923/j.nn.20150501.03

[87] Vijayanandan AS, Balakrishnan RM. Photostability and electrical and magnetic properties of cobalt oxide nanoparticles through biological mechanism of endophytic fungus *Aspergillus nidulans*. Applied Physics A. 2020;**126**:234

[88] Pavani KV, Sunil Kumar N, Sangameswaran BB. Synthesis of Lead Nanoparticles by *Aspergillus* species. Polish Journal of Microbiology. 2012;**61**(1):61-63

[89] Abdeen M, Sabry S, Ghozlan H, El-Gendy AA, Carpenter EE. Microbial-physical synthesis of Fe and Fe_3O_4 magnetic nanoparticles using *Aspergillus niger* YESM1 and supercritical condition of ethanol. Journal of Nanomaterials. 2016;**9174891**. DOI: 10.1155/2016/9174891

[90] Baskar G, Chandhuru J, Fahad KS, Praveen AS. Mycological synthesis, characterization and antifungal activity of zinc oxide nanoparticles. Asian Journal of Pharmacy and Technology. 2013;**3**:142-146

[91] Chatterjee S, Mahanty S, Das P, Chaudhuri P, Das S. Biofabrication of iron oxide nanoparticles using manglicolous fungus *Aspergillus niger* BSC-1 and removal of Cr(VI) from aqueous solution. Chemical Engineering

Journal. 2019a. DOI: 10.1016/j.cej.2019.123790

[92] Farrag MHM, Mostafa AMFA, Mohamed ME, Huseein MEA. Green biosynthesis of silver nanoparticles by *Aspergillus niger* and its antiamoebic effect against *Allovahlkampfia spelaea* trophozoite and cyst. Experimental Parasitology. 2020. DOI: 10.1016/j.exppara.2020.108031

[93] Rajan A, Cherian E, Baskar G. Biosynthesis of zinc oxide nanoparticles using *Aspergillus fumigatus* JCF and its antibacterial activity. International Journal of Modern Science and Technology. 2016;**1**:52-57

[94] Shende S, Bhagat R, Raut R, Rai M, Gade A. Myco-fabrication of copper nanoparticles and its effect on crop pathogenic fungi. IEEE Transactions on Nanobioscience. 2021. DOI: 10.1109/TNB.2021.3056100

[95] Yusof MH, Mohamad R, Zaidan UH, Rahman NAA. Microbial synthesis of zinc oxide nanoparticles and their potential application as an antimicrobial agent and a feed supplement in animal industry: A review. Journal of Animal Science and Biotechnology. 2019;**10**:57. DOI: 10.1186/s40104-019-0368-z

[96] Liu JM, Hu Y, Yang YK, Liu HL, Fang GZ, Lu XN, et al. Emerging functional nanomaterials for the detection of food contaminants. Trends in Food Science and Technology. 2018a;**71**:94-106

[97] Liu Y, Huang H, Gan D, Guo L, Liu M, Chen J, et al. A facile strategy for preparation of magnetic graphene oxide composites and their potential for environmental adsorption. Ceramics International. 2018b;**44**(15):18571-18518

[98] Zeng G, Chen T, Huang L, Liu M, Jiang R, Wan Q, et al. Surface modification and drug delivery

applications of MoS2 nanosheets with polymers through the combination of mussel inspired chemistry and SET-LRP. Journal of the Taiwan Institute of Chemical Engineers. 2018;**82**:205-213

[99] Zeng G, Liu X, Liu M, Huang Q, Xu D, Wan Q, et al. Facile preparation of carbon nanotubes based carboxymethyl chitosan nanocomposites through combination of mussel inspired chemistry and Michael addition reaction: characterization and improved Cu2+ removal capability. Journal of the Taiwan Institute of Chemical Engineers. 2016;**68**:446-454

[100] Nayantara KP. Biosynthesis of nanoparticles using eco-friendly factories and their role in plant pathogenicity: a review. Biotechnology Research and Innovation. 2018;**2**:63-73

[101] Jo YK, Kim BH, Jung G. Antifungal activity of silver ions and nanoparticles on phytopathogenic fungi. Plant Disease. 2009;**93**:1037-1043

[102] Abbas A, Naz SS, Syed SA. Antimicrobial activity of silver nanoparticles (AgNPs) against *Erwinia carotovora* pv. carotovora and *Alternaria solani*. International Journal of Biosciences. 2015;**6**(10):9-14

[103] Zakharova OV, Gusev AA, Zherebin PM, Skripnikova EV, Skripnikova MK, Ryzhikh VE,..., Krutyakov YA. Sodium tallow amphopolycarboxyglycinate-stabilized silver nanoparticles suppress early and late blight of *Solanum lycopersicum* and stimulate the growth of tomato plants. Bio- NanoScience. 2017. DOI: 10.1007/s12668-017-0406-2

[104] Ismail AWA, Sidkey NM, Arafa RA, Rasha MF, El-Bata AI. Evaluation of in vitro antifungal activity of silver and selenium nanoparticles against *Alternaria solani* caused early blight disease on potato. British Biotechnology Journal. 2016;**12**:1-11

[105] Kaur P, Thakur P, Chaudhury A. An in vitro study of the antifungal activity of silver/chitosan nanoformulations against important seed borne pathogens. International Journal of Science Technology and Research. 2012;**1**(6):83-86

[106] Saharan V, Sharma G, Yadav M, Choudhary MK, Sharma SS, Pal A.,..., Biswas P. Synthesis and in vitro antifungal efficacy of Cu-chitosan nanoparticles against pathogenic fungi of tomato. International Journal of Biological Macromolecules. 2015;**75**:346-353

[107] Dimkpa CO, McLean JE, Britt DW, Anderson AJ. Antifungal activity of ZnO nanoparticles and their interactive effect with a biocontrol bacterium on growth antagonism of the plant pathogen *Fusarium graminearum*. Biometals. 2013;**26**(6):913-924

[108] He S, Feng Y, Ren H, Zhang Y, Gu N, Lin X. The impact of iron oxide magnetic nanoparticles on the soil bacterial community. Soils Sediments. 2011;**11**:1408-1417

[109] Paret ML, Vallad GE, Averett DR, Jones JB, Olson SM. Photocatalysis: Effect of light-activated nanoscale formulations of TiO2 on *Xanthomonas perforans* and control of bacterial spot of tomato. Phytopathology. 2013;**103**:228-236

[110] Shenashen M, Derbalah A, Hamza A, Mohamed A, Safty SE. Antifungal activity of fabricated mesoporous alumina nanoparticles against root rot disease of tomato caused by *Fusarium oxysporum*. Pest Management Science. 2017;**73**(6):1121-1126

[111] Borchers A, Teuber SS, Keen CL, Gershwin ME. Food safety. Clinical Reviews in Allergy and Immunology. 2010;**39**:95-141

[112] Hoffmann S, Harder W. Food safety and risk governance in globalized markets. Health Matrix. 2010;**20**:5-54

[113] Pan M, Yin Z, Liu K, Du X, Liu H, Wang S. Carbon-based nanomaterials in sensors for food safety. Nanomaterials. 2019;9:1330. DOI: 10.3390/nano9091330

[114] Wu YN, Liu P, Chen JS. Food safety risk assessment in China: past, present and future. Food Control. 2018;90:212-221

[115] Duncan TV. Applications of nanotechnology in food packaging and food safety: Barrier materials, antimicrobials and sensors. Journal of Colloid and Interface Science. 2011;363:1-24

[116] Kerry JP, O'Grady MN, Hogan SA. Past, current and potential utilisation of active and intelligent packaging systems for meat and muscle-based products: a review. Meat Science. 2006;74:113-130

[117] Chaudhry Q, Scotter M, Blackburn J, Ross B, Boxall A, Castle L, et al. Applications and implications of nanotechnologies in food sector. Food Additives & Contaminants. Part A, Chemistry, Analysis, Control, Exposure & Risk Assessment. 2008;25(3):241-258. DOI: 10.1080/02652 03070 17445 38

[118] He X, Deng H, Hwang H. The current application of nanotechnology in food and agriculture. Journal of Food and Drug Analysis. 2019;27:1-21. DOI: 10.1016/J.JFDA.2018.12.002

[119] Mousa SA, El-Sayed SR, Mohamed SS, Abo El-Seoud MA, Elmehlawy AA, Abdou DAM. Novel mycosynthesis of CoO₄, CuO, Fe₃O₄, NiO, and ZnO nanoparticles by the endophytic *Aspergillus terreus* and evaluation of their antioxidant and antimicrobial activities. Applied Microbiology and Biotechnology. 2021;105:741-753. DOI: 10.1007/s00253-020-11046-4

[120] Komal R, Uzair B, Sajjad S, Butt S, Kanwal A, Ahmed I, Riaz N, Leghari SAK, Abbas S. Skirmishing MDR strain of *Candida albicans* by effective antifungal CeO2 nanostructures using *Aspergillus terreus* and *Talaromyces purpurogenus*. Materials Research Express 2020;7:055004. DOI: 10.1088/2053-1591/ab8ba2

[121] Omran BA, Nassar HN, Younis SA, El-Salamony RA, Fatthallah NA, Hamdy A, El-Shatoury EH, El-Gendy NS. Novel mycosynthesis of cobalt oxide nanoparticles using *Aspergillus brasiliensis* ATCC 16404—optimization, characterization and antimicrobial activity. Journal of Applied Microbiology 2020;128:438-457. DOI: 10.1111/jam.14498

[122] Abdelhakim HK, El-Sayed ER, Rashidi FB. Biosynthesis of zinc oxide nanoparticles with antimicrobial, anticancer, antioxidant and photocatalytic activities by the endophytic *Alternaria tenuissima*. Journal of Applied Microbiology 2020;128:1634-1646. DOI: 10.1111/jam.14581

[123] Atalay FE, Asma D, Kaya H, Bingol A, Yaya P. Synthesis of NiO nanostructures using *Cladosporium cladosporioides* fungi for energy storage applications. Nanomaterials and Nanotechnology 2016;6:28. DOI: 10.5772/63569

[124] El-Batal AI, El-Sayyad GS, Mosallam FM, Fathy RM. *Penicillium chrysogenum*-mediated mycogenic synthesis of copper oxide nanoparticles using gamma rays for in vitro antimicrobial activity against some plant pathogens. Journal of Cluster Science. 2020;31:79-90. DOI: 10.1007/s10876-019-01619-3

[125] Mahanty S, Bakshi M, Ghosh S, Chatterjee S, Bhattacharyya S, Das P, et al. Green synthesis of iron oxide nanoparticles mediated by filamentous fungi isolated from sundarban mangrove ecosystem, India.

BioNanoScience. 2019;**9**:637-651.
DOI: 10.1007/s12668-019-00644-w

[126] Yusof MH, Rahman AN,
Mohamad R, Zaidan UH, Samsudin AA.
Biosynthesis of zinc oxide nanoparticles
by cell-biomass and supernatant
of *Lactobacillus plantarum* TA4 and its
antibacterial and biocompatibility
properties. Scientific Reports.
2021;**10**:19996. DOI: 10.1038/
s41598-020-76402-w

[127] Al-Senani GM, Al-Fawzan FF.
Adsorption study of heavy metal ions
from aqueous solution by nanoparticle
of wild herbs. Egyptian Journal of
Aquatic Research. 2018;**44**(3):187-194

[128] Bozbaş SK, Boz Y. Low-cost
biosorbent: *Anadara inaequivalvis* shells
for removal of Pb (II) and Cu (II) from
aqueous solution. Process Safety and
Environment Protection.
2016;**103**:144-152

[129] Bradder P, Ling SK, Wang S, Liu S.
Dye adsorption on layered graphite
oxide. Journal of Chemical &
Engineering Data. 2011;**56**(1):138-141

[130] Brunet L, Lyon DY, Hotze EM,
Alvarez PJJ, Wiesner MR. Comparative
photoactivity and antibacterial properties
of C60 fullerenes and titanium dioxide
nanoparticles. Environmental Science &
Technology. 2009;**43**(12):4355-4360

[131] Darwesh OM, Ali SS, Matter IA,
Elsamahy T, Mahmoud YA. Enzymes
immobilization onto magnetic
nanoparticles to improve industrial and
environmental applications. Methods in
Enzymology. 2019;**1-22**. DOI: 10.1016/
bs.mie.2019.11.006.

[132] Almomani F, Bhosale R,
Khraisheh M, Kumar A, Almomani T.
Heavy metal ions removal from
industrial wastewater using magnetic
nanoparticles (MNP). Applied Surface
Science. 2019. DOI: 10.1016/j.
apsusc.2019.144924

[133] Zhang H, Chen S, Jia X, Huang Y,
Ji R, Zhao L. Comparison of the
phytotoxicity between chemically and
green synthesized silver nanoparticles.
The Science of the Total Environment.
2021;**752**:142264. DOI: 10.1016/j.
scitotenv.2020.142264

[134] Acharya P, Jayaprakasha GK,
Crosby KM, Jifon JL, Patil BS. Green-
synthesized nanoparticles enhanced
seedling growth, yield, and quality of
onion (*Allium cepa* L.). ACS Sustainable
Chemistry & Engineering. 2019;7:
14580-14590

[135] Sulaiman GM, Hussien HT,
Saleem MNM. Biosynthesis of silver
nanoparticles synthesized by *Aspergillus
flavus* and their antioxidant,
antimicrobial and cytotoxicity
properties. Bulletin in Matererial
Science. 2015;**38**(3):639-644

[136] Othman AM, Elsayeda MA,
Al-Balakocy NG, Hassan MM,
Elshafei AM. Biosynthesized silver
nanoparticles by *Aspergillus terreus*
NRRL265 for imparting durable
antimicrobial finishing to polyester
cotton blended fabrics: Statistical
optimization, characterization, and
antitumor activity evaluation.
Biocatalysis and Agricultural
Biotechnology. 2021;**31**:101908.
DOI: 10.1016/j.bcab.2021.101908

[137] Chen H, Luo L, Fan S, Xiong Y,
Ling Y, Peng S. Zinc oxide nanoparticles
synthesized from *Aspergillus terreus*
induces oxidative stress-mediated
apoptosis through modulating apoptotic
proteins in human cervical cancer HeLa
cells. Journal of Pharmacy and
Pharmacology. 2021;**73**(2):221-232

[138] Gupta I, Duran N, Rai M. Nanosilver toxicity: Emerging concerns and
consequencesin human health. In:
Rai M, Cioffi N, editors. Nano-
Antimicrobials: Progress and Prospects.
Verlag Germany: Springer; 2012.
pp. 525-548

[139] Gupta IR, Anderson AJ, Rai M. Toxicity of fungal-generated silver nanoparticles to soil-inhabiting *Pseudomonas putida* KT2440, a rhizospheric bacterium responsible for plant protection and bioremediation. Journal of Hazardous Materials. 2015;**286**:48-54

[140] Abu-Elghait M, Hasanin M, Hashem AH, Salem SS. Ecofriendly novel synthesis of tertiary composite based on cellulose and myco-synthesized selenium nanoparticles: Characterization, antibiofilm and biocompatibility. International Journal of Biology and Macromolecules. 2021;**175**:294-303

[141] Shaheen TI, Fouda A, Salem SS. Integration of cotton fabrics with biosynthesized CuO nanoparticles for bactericidal activity in the terms of their cytotoxicity assessment. Industrial and Engineering Chemistry Research. 2021;**60**(4):1553-1563

www.ingramcontent.com/pod-product-compliance
Lightning Source LLC
Chambersburg PA
CBHW081228190326
41458CB00016B/5718